EVOLUTIONARY THEORY AND HUMAN NATURE

EVOLUTIONARY THEORY AND HUMAN NATURE

by

Ron Vannelli
University of Central England (Birmingham), UK

KLUWER ACADEMIC PUBLISHERS
Boston / Dordrecht / London

Distributors for North, Central and South America:
Kluwer Academic Publishers
101 Philip Drive
Assinippi Park
Norwell, Massachusetts 02061 USA
Telephone (781) 871-6600
Fax (781) 681-9045
E-Mail <kluwer@wkap.com>

Distributors for all other countries:
Kluwer Academic Publishers Group
Distribution Centre
Post Office Box 322
3300 AH Dordrecht, THE NETHERLANDS
Telephone 31 78 6392 392
Fax 31 78 6546 474
E-Mail <services@wkap.nl>

Electronic Services <http://www.wkap.nl>

Library of Congress Cataloging-in-Publication Data

Vannelli, Ron, 1938-
Evolutionary theory and human nature / by Ron Vannelli.
p. cm.
Includes bibliographical references and index.
ISBN 0-7923-7473-8 (alk. paper)
1. Human evolution. 2. Human behavior. 3. Behavior evolution. 4. Social evolution.
5. Cognition and culture. I. Title

GN281 .V35 2001
599.93'8—dc21

2001038303

Printed on acid-free paper.

Printed in the United States of America

CONTENTS

PREFACE

So reason began to annihilate its enemies: God was dead; instincts were animal-like; emotions were condemned - they were enemies of reason, to be guarded against and strictly controlled. A strong will, strength of character, the pursuit of science - all children of reason - were to be admired, indeed worshipped. This was the gospel message of the Enlightenment; this was the road to modernity. But as reason began to worship itself, it began to deify itself. It became not enough in-itself; it 'needed' more; it became *Rationality*. And rationality cannot live with *irrationality*; it seeks to destroy it; but in doing so it kills the very essence of humanness.

The argument of this work will be that we must bring humanness back into our explanations if we are to advance our understanding of human behaviour. It will start with reason already having become rationality, explanations of human behaviour having desperately sought deterministic laws, social science having turned to teleology, social interactions having been formulated as systems/structures, and history having become progress; and while this was happening, emotions, self-interest, social ambiguity/disorganizaion and power politics had become evil - treated as reminiscent of 'the beginning', of disorder and chaos, products of an age 'before reason'.

This will be a story, however, which might have consoled Charles Darwin, who had come to fear that he was the 'Devil's Chaplain' for advocating the scientific importance of non-teleological explanations which included elements of *chance, conflict* and *non-progressive* history. This is because the approach developed here will try to rehabilitate emotions, self-interest, social ambiguity and power politics; it will, in a sense, argue the 'Devil's' case; or perhaps more accurately it will seek to reunite Enlightenment gods with their demons. It will try to bring reason and emotion back together and it will attempt to put self-interest, social co-operation and personal and social politics into the same framework; it will seek to demonstrate the role of a human *'will to power'* and the importance of practical politics in this process. The notion of a grand design will be conspicuous by its absence.

Human nature will loom large in this undertaking. It will not be a deterministic human nature, nor will it be a rational one capable of continually learning and perfecting itself. Rather, it will represent the dynamic proesses in which our species typical emotions and reason struggle with each other. It will not depict a human nature with a mission or known direction, but a set of mental processes which are the product of natural and sexual selection - with both their chance and 'order in process' elements. It will not be a search for 'Truth'; it will seek to understand and explain.

The first major contention, as suggested, will be that we must take on board much of the 'Devil's' contribution to our view of humanness if we are to have a chance of understanding ourselves *as we really are*. There has been, and will be, resistance to such a view; after all it does not present an *ideal* view, it is not even (on the face of it at least) a very 'nice' picture of humankind. Unfortunately, ideal or nice pictures are not always the best representations of something, nor do such pictures necessarily cause their subjects to become so. If the Devil's view is less distorting than any alternative picture, and our goal is understanding, we have no choice but to embrace the understanding it might offer.

And a major assertion of this work will be that not only must we consider the Devil's view, but that it is, in fact, less distorting than the majority of favoured alternatives and is also quite an accurate representation of humanness (and, incidentally, not necessarily as negative an interpretation as so often imagined). I will start by looking at some of the specific distorting *consequences* of deifying reason - and very significantly, at some of the less than desirable consequences for humanity of having done so. Possible alternative conceptual, theoretical and empirical approaches to those which have deified reason and embraced teleology will be examined, as will some of the 'moral', conceptual and theoretical issues raised when we set out to link biology and the social sciences using a Darwinian perspective (Chapters 1-3).

This will be followed by a look at a wide body of evidence from the neural sciences and various branches of psychology in order to suggest an evolutionary based, non-teleological model of human nature (Chapters 4-6). This model is to be made up of potential bioelectric/chemical, emotional/ psychological and cognitive elements. In the Devil's dynamic – but often unsettling – picture of human nature, emotions, desires and fears play as important a role as human reason; love and hate have their place, the 'will to power' is ever-present. Cognition is not only about solving complex probems but also about evaluating, stereotyping, distancing and deceiving (both self and others).

Based on the theoretical argument and model of human nature developed, it is then suggested that human behaviour can be classified into 'natural patterns', or 'natural kinds' based on 1) a high degree of coherence derived from identifiable human nature propensities, and 2) on their empirically discovered anthropological and historical universality (Chapters 7-8). It is for these reasons that a non-teleological social science is possible; one clearly dependent upon a full understanding of the working of human nature and on a study of history.

As noted, any notion of a grand design (God or Nature, no matter) as a driving force in human history will have to be abandoned. Banished also from our theorizing must be the common human bias that human nature was *necessarily* made for being perfected (or for perfecting its environment), that it is characterized by a super-rationality which lifts humans above any remnants of an animal nature, and, above all, that humans have a larger, universal purpose – such as to take us to utopia, for example. This development will not be easy. Humans are not quick to give up the notion of a higher purpose, even for the prospect of a science of humans; indeed, especially not for the prospect of a science of humans. Nevertheless, even though the search for knowledge is often uncomfortable, human curiosity is such that, in the end, it cannot be denied.

Acknowledgments

This work has had a long gestation. Many of the people who should be thanked are not easily identifiable. Students, colleagues, friends, relatives and some complete strangers said things which ignited something, which upon reflection (often no doubt unconscious reflection), was used. Numerous scholars and teachers, whose work has been very influential but, perhaps because of the long time since their ideas were absorbed (and likely not always correctly), have not in every case been properly recognized in the referencing. So if it were possible I would thank them now, and apologize for my forgetfulness or misconstruction of their ideas. In terms of names, however, I will have to confine recognition to those who did something memorable (and for me, special) to help this project along.

Vic Heatherington read an early version and made many useful observations, not the least of which was to point out that my (at that time) somewhat fear/danger oriented look into the nature of humankind could benefit from an equal emphasis on a search for human desires. Harry Bauer saw the challenge that this same early version represented and encouraged me to keep on with it; Fred Willhoite, too, was concerned that it would rankle, but encouraged its continued development. As the work progressed, and was revised, Alan Rowan and Judy Tweddle spent long hours discussing what is now Chapters Four, Five and Six and how we might 'draw' them. Although the resultant diagrams are no fault of their's, their thinking was very much part of the process of getting to what is there now. Pat Needle read various versions of the manuscript in order to try to make sure the grammar, syntax and general presentation was not too far out; she also made many valuable academic points. Later on, Malcolm Hughes and Brenda Shute read what are now Chapters Four, Five and Six and made helpful comments. Bill Roper read a number of sections concerning the impact of Darwin and raised a number of points I had to take on board.

Kate O'Shea saw great value in an interdisciplinary approach to scholarship and teaching which went back to the question - 'what is it to be human?', and encouraged me in a number of directions; and her students contributed much, unknowingly, to the final product. Robert Kornreich, too, engaged with many of the ideas put forward here during episodes of team teaching, discussion and seminars; although he did not always agree with various notions, he understood their intent and reasons and was always ready to engage in scholarly conversations or debates, as the circumstances dictated. And he sincerely enjoyed seeking answers to the puzzling questions which have engaged theologians, philosophers and scientists for a very long time.

Paul Waddington had a great deal of faith, and understood the aims, and also looked for answers to big questions, and never seemed to doubt that the project was worthwhile and that the questions being asked should be asked; although not always in agreement with my views, his encouragement was often just what was needed when things were not going well. He also read various versions of possible 'starts' of the book, making important stylistic and academic points. As the project neared its end, Malcolm Hamilton and John Richardson read the entire manuscript, gave encouragement, made suggestions and consulted over the exact form it should take. None of the above, of course, would agree with me on everything but from my point of view this makes their contribution all the more valuable because I had to decide why I would stay with my interpretations on those occasions where I did not follow their advice (although often I did follow their advice). Near the end, Linda Harland took a final look and made some important proofing corrections.

Most of all, I must thank my wife, Susan Vannelli, who struggled with a number of rewrites; always sharpening up the presentation, teaching me commas, questioning the validity of many points, suggesting new and better ways of expressing certain things; postulating ideas; but above all, believing in the scholarship it represented, and in my determination to complete it, even when I seemed to always be changing things; so, despite often despairing at what seemed to be yet another re-write, she stuck with it, and me. By the end she knew its argument as well as I do. This book is dedicated to her.

CHAPTER ONE
IN SEARCH OF HUMAN NATURE: REASON AND RATIONALITY

As human beings, we like to think that there is purpose in our lives. We commonly assume that there is aim and direction in what we do. Many of us feel that we are part of a greater design. In a sense this is good because it makes us feel good. It concentrates our minds on our goals and gives us pleasurable feelings of success when we achieve them. Unfortunately, this underlying belief in, and search for, an ultimate design and purpose is pure teleology. The consequence of this teleological quest is that it does not take us beyond an 'intuitive sense', 'idealistic hope' or a 'God's will' explanation of human behaviour and history. However, when couched in terms of a search for *universal rationality,* it seems to.

REASON AS RATIONALITY

What has happened is that human reason - more accurately reasoning - has become *Rationality.* Reasoning involves those processes of mind we refer to as imagining, experimenting, reflecting and thinking. As a means for understanding it implies no special design or purpose in that which is being explained. *Rationality,* on the other hand, pictures the universe as operating according to universal *laws* and *structures* which can be discovered by *pure* reason and/or science. It includes a notion that when these are discovered, 'efficiency' and 'progress', if not 'goodness', 'happiness', justice and universal morality, will soon follow. Advocates of Rationality generally hold that human emotions, desires and fears are major obstacles to our ability to discover the 'true' nature of the universe and, therefore, hinder our ability to achieve any of the above types of 'progress'. Unfortunately this approach has led numerous philosophers and, more recently, social scientists and policy makers into a variety of arguably very misleading propositions and explanations concerning human nature and 'the human condition'.

In the case of social scientists (see note one p., 234), for example, direction, design, purpose and cause are often attributed to certain human propensities: for instance, propensities for generating abstract ideas/beliefs, for co-operation, for being social, for 'altruism' and for accumulating knowledge. This has frequently taken the form of attributing *Final Cause* to such things as, for example, 'Rationality', 'Culture', the 'Division of Labour', 'Sociability', 'Social Systems', 'Science/Technology' and 'Progress'. This has more often than not also led to spiritual explanations in that these, and commonly some of their constituent parts, are given the status of a *higher* essence.

Furthermore, the teleology of attributing design, purpose, supra-human causal power and a sense of inevitability to the above essences can suggest that they have *needs.* The concept of needs generally implies a *moral necessity* to fulfil them (Cf., Kant, 1974; for discussions see Fitzgerald, 1977a; Plant, 1991, Chap. 5). Thus we reach an even higher spirituality, that of attributing *moral* value to these essences. So, for example, in the social sciences it has been argued that we need stable societies, and stable societies need functional families, and so functional families become not only necessary but also, in the minds of many, moral requirements; or that we need justice and that equality is necessary for justice and so equality becomes a moral necessity.

Additionally, a great deal of 'modern' thinking tends to reinforce the spiritual dimension of this teleology in its 'analysis' of the place of human *reason/rationality* in the universe. Humans, the argument usually goes, are a *final* result of a greater *rational* design because the human capacity for reason *must* have been specially produced to understand it: ' why else'– is the view –'would human reason exist if it did not have such a special purpose?'. Humans thus have attributed to reason/rationality, and consequently to themselves, a clear superiority over nature.

The human tendency to spirituality and teleology has also suggested to various philosophers, social scientists and policy makers (see note one, p, 234) that, among other things, there is a desirable *end* towards which human civilization is *necessarily* moving. For some thinkers utopia is not only possible but also discovering its characteristics, and thus helping to achieve it, is a *major* part of the greater human (moral) purpose. History, then, is *progress*. And individuals do not have to feel that they will be un-rewarded slaves in its inevitable achievement; they are the very reasoning force behind it Moreover, improvement in everything significantly human - ranging from sex (Cf., Masters and Johnson, 1970) to the condition of the human psyche (Cf. Kelly, 1955; Maslow, 1987) and to the social formations and the 'world community' in which we live (Cf., Rousseau, 1968; Kant, 1963, 1974; Saint-Simon, 1964; Owen, 1813) - has come to be considered not only within the bounds of possibility but often a human duty to strive for. Thus, it is felt, it is the moral duty of individuals (and of societies) not only to participate in progress but to do so with enthusiasm and joy. And this need not be difficult because now social science will be leading the way, enlightening and clearing from the path 'superstitions' and theologies which had hitherto been stumbling blocks scattered on the road to human social progress and individual perfectibility; happiness has become a human 'right'.

As noted in the Preface, it is my argument that we must escape this recreated teleological spirituality (from earlier theological versions) concerning human existence if the social sciences are to have a chance to understand humans and human social life *as they are*. Developing a non-teleological approach to human nature, human society and human social evolution is essential for this task, and the way to do so is through an empirically based science. This should be possible so long as we are not too restrictive in our definition of science. That is, if we consider science as a *constant process* of observing and theorizing in a *non-teleological, non-spiritual* manner we will be able to shed important new light on the human condition, including the human tendency to spirituality.

But what is wrong with spirituality and teleology? Those who pursue such ends as have been described above, for example, have generally been concerned with the good of humankind; they have wanted to uncover *truth*, to be able to prescribe a better way. And no less a thinker than Aristotle found teleology a very useful way of striving for 'The Good' and for understanding and explaining what appears to be functional inter-relationships within human societies. Many thinkers since have theorized both practical and theoretical activities on the basis of privileging the clear human ability for 'instrumental (means-ends) rationality' and for creating 'functionally integrated' social organizations. It can be argued, moreover, that the human capacity for conceptualizing design, purpose, and for striving for utopia, has been the driving force behind human civilization; that it is this ability which makes humans more than animals.

Fair enough. Certainly a great deal of social knowledge, useful organizational procedures and workable social policies have come from this approach during its development into the Enlightenment-inspired social sciences ('the classical tradition'). In so far as we are natural teleological thinkers (see Chapter Six) we have often been able to identify relevant means *once* goals have been decided upon. This has included the generating of generally appropriate organizations, hierarchies, job descriptions, criteria of achievement, plans and agendas for a number of activities, including those involved in survival and reproductive pursuits. There is no doubt that 'instrumental rationality' has served humans well in identifying *means* and in deciding on reasons why given means are not achieving specific goals.

This is fine. The problem comes into focus when we try to explain why and how specific goals are decided upon in the first place; or when we want to know why a concocted teleological functionality does not explain a great many things - or if it does, why we are suspicious of the circularity of the answer (for example, the Holocaust being explained by the 'needs' of German society at the time - usually more eloquently expressed than this). 'Good vs. evil' answers do not take us at all further. In the end, the teleological and spiritual propositions concerning societies and history discussed above are nothing more than assumptions about human motivations and human nature generally based on postulated *purposes*. Very dubious assumptions at that. For example, historical evidence certainly does not make it clear that civilization has been an unfolding of some master plan or design. The role of chance, luck, sometimes of chaos, cannot be ruled out; and based on observations of history, indeed, often seems to loom large.

Furthermore, it is far from clear that the concept of rationality has any meaning separate from describing humans attempting to set goals and to select relevant means for specific ends and as a process of *rationalizing* after the fact. Even this limited usage, besides not considering the multitude of emotional and cognitive processes which are involved in the difficult task of deciding what are 'relevant' means, as noted, leaves out the extremely important problem of understanding how and why humans come up with specific goals in the first place.

In other words, a major problem with the traditional social sciences is that their teleological nature has meant that they tend to be circular, if not tautological. They most often are unable to answer really difficult questions concerning causes; there is a tendency to assume the answer in the questions posed. If, for example, we assume that humans have specific goals (rationally derived), we can take all of their behaviour as being aimed at fulfilling them; if these behaviours do not seem to fulfil these goals, we usually do not question our theories of human motivation or causation within societies but rather argue that a particular individual is not acting *rationally* or that there are exceptional conditions. If we decide that healthy societies *need* stable families we have a social explanation for the relative stability of family life when we find it. Given our assumption about the need for stable families we do not have to take the existence of non-stable families as a refutation of our explanation concerning the nature of society; we can label such families as deviant, pathological and/or as evidence of society in trouble. And so even our theory of stable societies requiring stable families seems supported by the existence of unstable families!

Moreover, it does not seem that humans have always been well served by 'rationality' and 'morality' or that they have brought utopia, or that they are taking us in that direction (even if we could agree the nature of such a state). That is, history seems to reveal that as much, if not considerably more, harm has come to humans from attempts to put rationally conceived, utopian/moral systems into practice, or from trying to fulfil some 'rationally' postulated historical purpose or other, as from all the muddle-through, semi-disorganized, contradictory, gut thinking, hedonistic, lust motivated behaviour typical of the so called rationally unsophisticated.

One need only consider a few examples to make this point vividly. The Communist Party in the Soviet Union destroyed millions of human beings (Conquest, 1971) in its failed attempt to establish what its members considered to be the ultimate in a rationally devised moral socio-political/economic system. Zygmunt Bauman (1988, 1989) has shown how the Nazi destruction of millions of Jews, Slavs and other 'less than human types', was done with the use of considerable political, bureaucratic and technological 'rationality'. His argument, in fact, is that in many ways these forms of 'rationality' made the Holocaust possible. Indeed, it has been argued that the Second World War itself was a result of excessive Enlightenment inspired rationality and moral enthusiasm on all sides (Morgenthau, 1946). Certainly, the Vietnam War was generated and perpetuated by individuals considered to be 'the best and the brightest'. They were perceived to be the very essences of rationality (as opposed to politics) and saw themselves as such (Halberstam, 1969). Few activities have generated as much scientific rationality, applied with such extreme zeal, as the creation of weapons of mass destruction (Rhodes, 1986, 1995; Read, 1994). It is important to note that in all of these examples, *rationality* was often preached with the zeal usually found in messianic religions; it was often also identified with morality.

There has, in fact, been a long history of scholarly work (not always connected) which seems to have, more or less, recognized the dangers of teleology and of deifying reason. They have recognized problems for both developing a scientific understanding of humans and for avoiding the *political dangers* of not recognizing emotions, power, politics and 'irrationality' as basic motivators of human behaviour. For example, the works of such scholars as Machiavelli, Hobbes, Voltaire, Ferguson, Hume, Darwin, Nietzsche, Weber, Freud, Morgenthau and Foucault, have more or less challenged, each in their own way, most of the spiritual and teleological propositions of the classical tradition. Indeed, it can be argued that in many ways their work challenged several of the key assumptions of the Enlightenment project itself.

Realism And Rebellion: A Critique Of Enlightenment (*Rationalistic*) Spirituality

A number of the conceptual developments and theoretical concerns of the above are so central to the tasks and approach of this work that it is worth looking more closely at some of the key ideas and concepts which emerge from or are strongly emphasized in their work. It is also interesting to note that all these writers have more or less been demonized by the traditional social sciences (and also by theology, much of philosophy and large segments of public opinion). It is tempting to designate this group of thinkers, 'the realist tradition'. 'Tradition' might be misleading in that these writers generally did not use each other's work as the basis of building a paradigm. Nevertheless, the earlier writers greatly influenced the later ones and there is a considerable similarity of thought to be found in their work. The term 'realist', on the other hand, seems more appropriate. In the first place, they shared with philosophical realism - and with the classical tradition in the social sciences for that matter - a very strong belief that there was a real world out there to be studied and that human reason (thinking and reflecting) and science (observing, theorizing and questioning) were real ways to study and understand it. In this regard they are firmly in the Enlightenment tradition.

A classic example of this belief in empirical science was Darwin's almost obsessional fear of publishing until all possible observations and measurements had been made, and his belief that sexual and natural selection could only be established with substantial evidence from geology, palaeontology, animal behaviour and observations of reproductive competitions. His powerful desire to create a *non-teleological* theoretical explanation that united geology, palaeontology, biology, chemistry and animal (including human) behaviour makes him a strong representative of science in a broader, theoretical sense (Darwin, 1859, 1872, 1874, 1950).

For their part, Machiavelli (1950, 1963), Hobbes (1968), Voltaire (1980, 1972, 1963), Ferguson (1966), Hume (1963, 1968, 1975), Nietzsche (1886, 1887, 1887a, 1974) and Weber (1978) were all convinced that any understanding of human society required *detailed studies* of comparative history, especially of political history. Each was concerned to scientifically discover the human nature processes and political conditions and 'principles' from which state formation, or at least 'civil society', moralities and forms of governance, emerged. They were equally keen on the use of comparative psychology, anthropology and sociology to discover 'facts' and to theorize about human motivations, status concerns, power, politics and the possibility of a non-teleological evolution of civilization.

Freud was quite insistent that the unconscious had to be established through detailed studies of dreams, jokes, slips of the tongue, fantasies, universal dramatic stories, ancient mythology and psychoanalysis, all sources of evidence which are within the grasp of science (1971). He also had a strong desire to theoretically link evolutionary biology, physiology and psychology for an understanding of consciousness (Cf., Freud, 1971, Lecture XXXV, 1976, 1940). Foucault (Cf., 1979, 1980, 1980a) used history and organizational case studies to establish his theoretical notions concerning control and power relationships. He even sometimes claimed to be an ". . . empiricist: I don't try to advance things without seeing whether they are applicable" (Foucault, 1988). Voltaire (1972, 1980) thought Newton to be a rare genius and was very impressed with Locke's empirical methods. He did a great deal to popularize and advocate the ideas of both of these on the European continent (Gay, 1967, 1969, 1988).

A second reason for considering these thinkers realists is that they not only rejected, often with considerable vigour, the spiritual and teleological assumptions and approaches of religious thought, but also much of philosophy and, latterly, the classical social sciences. For example, the early thinkers ignored or largely rejected theological Final Causes such as God, 'The Kingdom of God' (as seen in the Scriptures), The Word of God (spoken through the Church) or forces such as Good and Evil (from a variety of official and local sources). This tradition of skepticism was maintained by the later realists. Having seen the Church and absolute monarchs defeated, they remained dubious, if not out and out rejecting, of notions such as 'The Absolute', 'Society', a 'Collective Consciousness', 'Social Systems' (including 'capitalism' as a system or 'the proletariat' as a historical force derived from it), and, above all, 'Progress'. In other words, they did not embrace most of the grand abstractions and optimistic projections for humankind which were evolving with the emerging social sciences, considering them as being from the same stable as the grand religious and philosophical teleologies.

This began a rift with certain enlightenment ideas which has not been resolved to this day. It was not simply that the above thinkers were rejecting piecemeal those things which they did not like about the emerging societies around them but, rather, they appeared to be rejecting everything that was being deemed positive at the time. Their extreme skepticism concerning a universal rationality, design and teleological purpose, for example, hit at the very heart of the notion of a progressive nature of history, so common and precious to many Enlightenment theoreticians, and certainly to the developers of classical social theory. History, both natural and human, for realists was to a significant extent a chance, although not completely random, process.

Darwin's great revolution was to postulate natural and sexual selection as a process which occurred separately from any postulated or predictable outcome or purpose (above references for Darwin). However, this process, nevertheless, had identifiable principles which were relatively consistent. For example, certain variations in reproductive outcomes (now known to be largely because of genetic mutations) occurred randomly, as did many environmental changes. But there was a consistency to be found in the powerful tendency of biological characteristics to *usually* be replicated almost exactly (now known to be because of the nature of genes), and for what already existed to greatly affect the degree and manner in which any variations (mutations) were or were not successful (see also, Dawkins, 1976, 1997; Cronin, 1991; Foley, 1996).

When considering human history, Machiavelli (1950), for example, considered that "Fortune" was responsible for a very significant amount of it; and its power was not to be ignored, because ". . . men may second Fortune but not oppose her. . ." (p., 383). Nevertheless, there were discoverable principles of history based on *universal human passions* and the fact that history could only build on what was already there (see also, Hume,

1968). These principles could, to some extent at least, be used to maintain and govern a good principality.

> ". . . for where men have little wisdom and valor, Fortune more signally displays her power; and she is variable so the states and republics under her influence also fluctuate and will continue to fluctuate until some ruler shall arise who is so great an admirer of antiquity as to be able to govern such states so that Fortune may not have occasion, with every revolution of the sun to display her influence and power" (pp., 387 - 388; on the role of 'Fortune' see also Nietzsche, 1886, 1887, 1968; Ferguson, 1966; Weber, 1981; Freud, 1961,1971 - Lec., XXXV; Hume, 1964)

For these scholars, emphasis varying from individual to individual, the real world was a more or less largely unpredictable constant struggle for status, pleasure, survival, self-respect and for both the control of passions and self's circumstances (a 'will-to-power'). It was about sexual, personal and practical politics, sexual selection and natural selection; it was about the evolution of human consciousness and civil or political society. It was not a process that was necessarily going anywhere special; above all it was not a teleological process of inevitable historical progress.

Being at odds with the mainline Enlightenment's legacy of a universal rationality and purpose, these writers accorded human emotions ('passions') a prominent place in the realm of human motivators. While accepting the existence of human reason, and the importance of science, passion was often considered to be stronger than reason: ". . . our intellect is a feeble and dependent thing, a plaything and tool of our impulses and emotions. . ." (Freud, 'letters', cited in Hughes, 1959, p., 143, see Freud, 1961; Hume, 1968). And in some cases it was even argued that passions were more worthy than reason; "Reason is and ought only to be the slave of the passions" (Hume, 1968, p. 415; 1968, Book II; see also, Voltaire, 1747; Nietzsche, 1990, 1968 - but see 1961). That is, who could argue that reason and rationality were superior to love, empathy, sympathy and a will to live or to express self (a 'will to power'), for example, in the formulation of, say, systems of justice? And there is a real sense in which we must keep our eye on those who claim a monopoly of rationality lest they create too many World War IIs (Morgenthau, 1946). Foucault (1988) sees continuing employment for philosophers here because ". . . the role of philosophy has also been to keep watch over the excessive powers of political rationality - which is a rather promising life expectancy" (p., 58).

This brings us to a further reason for describing these thinkers as realists. As noted, they firmly believed that there was a real world out there to be studied. Therefore, despite their rejection of a universal rationality, design, purpose and final cause, they shared a desire to find an alternative universal of human existence which could become the basis of a human science. This usually took the form of a search for either 'human nature', or at least a universal essence of human nature.". . . the Science of Man is the only solid foundation for the other sciences . . . [and]. . .human nature is the only science of man" (Hume, 1968, pp, xx and 273). This human nature had to include both the human ability for reason and human passions (emotions). But, above all, it had to be based on what these thinkers considered scientifically discovered, real aspects of the world. It was generally made clear, moreover, that it had to be discovered independently of any spiritually or teleologically postulated design or utopian purpose.

Freud, for example, in searching for the essence of human consciousness, often tried to explain psychological processes in biochemical terms as they might have evolved through processes of natural and sexual selection (Freud, 1891,1895, 1940), see also; Ritvo, 1990; Gay, 1990; Hughes, 1959). Nietzsche wanted the will to power to be considered a non-teleological, scientific concept based at least on psychology (Nietzsche, 1968, 1886, 1974), which had eventually to be explained in physiological terms if not more directly as some force of universal energy describable in physical terms (Nietzsche, 1889, 1961, 1968; see also, Hollingdale, 1973, p., 58; Conway, 1996; Clark and Leiter, 1997a). The others in our list of realist scholars generally searched for universal 'passions' or emotions which centred around human concerns for security, self-identity, status, honour and control of self's circumstances.

Realists, in other words, searched for an essence, or essences, of humanness which did not have a teleological basis, but ones that would also allow them to escape total relativism. Thus, as we have seen, history – natural and social - according to them, was to a significant extent due to chance but was not completely random. It would have discernible forms based on a universal human essence (nature) and its own momentum. That is, history could only build on human nature, including the human ability for reason, and/or what existed before and so would have discoverable patterns. This would be the case regardless of the random genesis of specific motivators - internal or external - for change and a lack of overall design, purpose or direction.

And, most emphatically, as noted, teleology was decidedly rejected, as was the notion that humans represented a higher spirituality; everything was not necessarily, as Dr. Pangloss (representing Leibnitz) had proclaimed, 'for the best in the best of all possible worlds' (Voltaire, 1759), and each event might have a cause but it was not necessarily a cause towards perfection (Voltaire, 1747), or, indeed, toward any direction at all.

On the surface at least, the notion of realism might seem strange given that these writers were searching for a human nature (essence) which included: a concern for human 'passions', 'the unconscious' and/or a 'will to power'. On the other hand, almost everyone accepts that these concepts represent something that is real. Love, empathy, hate, envy and jealousy, for example, represent passions we know exist. And we have thoughts, feelings, anxieties, fears, dreams and memories which we can not easily articulate, but which suddenly seem to overwhelm us. We, in other words, often experience fears of losing control to something powerful and deceptive within ourselves. At the same time our delight in mastering our environments, our feelings of powerful love, our sometimes great senses of achievement, our thrill in winning, our desire for revenge, all testify to 'the will to power'. These all represent something experienced as real. And this cannot be so easily said for such things as 'universal rationality', 'social systems', 'the proletariat', 'class' or 'progress', for example.

Furthermore, if the processes described by the concepts of 'passions' (emotions), the unconscious and a will to power are difficult to observe, the more we must place our hope in reason and science, argue the realists. While these provide no magical way to discover 'truth' they are better, especially when used together, than any other method we have of understanding (obtaining knowledge about) nature, including human nature. Reason may not be stronger than passion, but we must continue to use reason to slowly accumulate knowledge if we ever hope to be free, argued Freud (1961, pp., 96-97; Gay, 1954, 1990; Hughes, 1959) - disagreeing somewhat with Hume, and perhaps also with Nietzsche, about the virtue of passions, but not about their strength or the importance of science. Moreover, many of these scholars expected that science would one day uncover the physical nature of some of the key essences; as noted, for example, Freud and Nietzsche believing that the unconscious and human will to power would, hopefully, at some point be explainable in some form of biochemical terms.

Some might argue that a better name for this group of scholars would be 'essentialists', and certainly many of them have been so labelled in the past. I think, however, that is misleading. First, the essences searched for were not claimed to determine behaviour in a way that a notion of essentialism usually implies, but rather, to serve as concepts which represented the general motivational forces and parameters in which behaviour was enacted. Human consciousness was not deterministic but contained numerous passions (motivating and restricting) which would be more or less activated depending on the life history and circumstances in which individuals found themselves.

Indeed, in Freud's account, for example, the unconscious was often at war with itself, not guiding humans to predetermined outcomes but rather subjecting them to emotional struggles, not uncommonly pushing them in different directions at the same time. In all the accounts, human reason and human passions were often in direct conflict, sometimes in a sort of love/ hate relationship. The will to power took many forms depending on the history, circumstances and 'Fortune' of a particular group of individuals in their struggle for security amid the vagaries of practical politics. Rather than setting up essences to become deterministic forces in human behaviour, realists were looking for essences which got rid of the distinction between human and nature, between human and animal, between reason and emotion, between moral and immoral and between good and evil. Humans were part of nature, and that was that, and causes of behaviour were many and varied, just as they are in nature.

A final reason why these thinkers might be called realists is that many of them have long been so labelled in the tradition of political theory. This is because within that tradition they made the study of *power* and *power politics* of primary importance. They gave struggles for power and politics (personal and social) a life of their own, not exactly independent of rationalistic, ideological, economic, organizational, technological and/or historical events, but as human nature based driving forces for dealing with all these. This is most vividly seen in Machiavelli, Hobbes, Voltaire, Hume and Weber in their analysis of the political processes involved in state/government formation.

Weber (1978), for example, had specifically theorized the importance of power and politics when he suggested, following Machiavelli, that a state could only truly be a sovereign state when its leaders (government) had a monopoly of the capacity for physical violence. This, however, for Weber, tended to generated a struggle sufficiently costly that leaders were always in the *political* process of trying to legitimize their authority through other means. All of this is real because both physical struggles and the political processes involved in legitimization were made up of real human actions and real humans were always in the process of giving meaning to those actions.

This approach to power and politics, however, was most specifically labelled as 'realist' by Hans Morgenthau in the field of international relations where he describes "realist theory" as being the study of ". . .interest defined as power" (Morgenthau, 1954, p., 5). His view was that, while a wide variety of factors would go into defining national interest, once decided upon leaders of states would, inevitably, use whatever means, strategies, techniques, resources or advantages they had at hand to protect the state's interests. This was because, there being no higher authority to call upon in international relations, humans can only be safeguarded by having power. If we change the concept of 'national interest' above to ego, self, group, factional or professional interests, we have a relatively good picture of how this entire group of realist scholars generally approached the notion of power. Indeed, Morgenthau (1954) himself noted that

> "In view of [the] ubiquity of the struggle for power in all social relations and on all levels of social organization, is it surprising that international politics is of necessity power politics? And would it not be rather surprising if the struggle for power were but an accidental and ephemeral attribute of international politics when it is a permanent and necessary element of all branches of domestic politics?"(p., 31).

Power is arguably the master concept in the work of realists - except for those who put reproductive competitions or sex before power; unless, of course, we consider power relationships to be sublimated sexual/reproductive competitions, in which case they become the same. Whatever the case, power for realists was not a thing with magical qualities but rather a description of processes in which self-interest, 'creative subordination' and domination, claim and counter-claim, leading and following and legitimizing and de-legitimizing were common; and sometimes violence was used in these claims.

Power in their work, then, was a concept which described practical, strategic, shifting, unstable, *real human politics*. It represented real human activities, ones based on human nature and human social interactions. It was not a *thing* with a magical quality which, in itself, gave control or domination. It was not a struggle between master and slave. The 'will to power' could manifest itself as a struggle for 'self-becoming', as a 'will to truth', as a 'will to falsity', as 'seduction', as a 'will to delusion', as a battle for professional control or as strategic games to get one's, or one's group's, way. At a more social level it could be manifest as conquest but also as 'balances of power' or "agonistic democracy" (Hatab, 1995;). In other words, power represented, for realists, a number of human reactions, strategies, mechanisms and techniques for continuing to be human.

Above all, power was not an *ingredient* of a social system which helped structure that social system in a particular way. It was not something which particular human social locations afforded. Human 'societies' were the result of human politics. Anything approaching 'a social whole' was a product of politics not the other way around; 'a whole' was a network of power relationships. And wherever one was in this network one could not escape the pleasures or dangers of playing the power game. Power relations were never harmonious, almost always confrontational, but not necessarily in an unpleasant manner.

Thus, in this approach, power represents the linking of relationships from the level of self to the level of international relations. It is not something which can be reduced to a few variables. It is not always easy to theorize about. But it is, for realists, of extreme importance in understanding the human condition. As noted, it might even be considered the key concept of this group of thinkers. Indeed, for some of them, power seemed almost to transcend human actions. But this transcendence was not as an essence or a 'quality' but rather as a pattern of behaviour which included a very easy acceptance of customs, patterns of behaviour and even subordination. In Hobbes' perspective, for example, once humans actually gave up individual sovereignty to the monarch in order to be secure, the monarch (the state) in essence gained a 'reserve of power'. This was because the will of the people had been translated from an active voice to a passive voice. Nevertheless, this could never be made completely secure as human passions could always get out of hand (Hobbes, 1968; see also, Clegg, 1989).

Hume suggested that people often obey a government out of 'habit' (in Stirk and Weigall, 1995, pp. 52-53), clearly a 'store of power' as far as government officials might be concerned, but habits can be very fickle indeed. Weber forwarded the notion that an ability to control circumstances and others based on force or charisma would often be 'stored' as 'traditional authority' or legitimized as 'rational-legal authority'; but these were continually problematic and evolving. For Nietzsche and Foucault political actors were primarily products of networks of power (Cf., Simons, 1995; Best and Kellner, 1991).

> "Power . . . is never localised here or there, never in anybody's hands, never appropriated as a commodity or piece of wealth. Power is employed and exercised through a net-like organization. And not only do individuals circulate through its threads; they are always in the position of simultaneously undergoing and exercising power. They are not its inert or consenting target; they are always also elements of its articulation. . . individuals are the vehicles of power, not its application" (Foucault, 1980, p., 98).

The notion of a 'store of power', then, does not mean that power is a magical or supernatural force outside human interactions. It means that humans are born into social relations which are, in essence, power networks; by instincts and by learning humans play power games long before consciousness is fully developed. It means, following Aristotle, that humans are political creatures; and it means, following this, that power is a key concept in understanding their social life.

Realists And The Classical Social Science Tradition

Whether we will be allowed to call these thinkers, and numerous others like them but less famous, 'the realist tradition', remains to be seen (certainly each, in their own way, did not strictly adhere to all the characteristics described above as representing realist thinking). I doubt very much that most of them would have liked being called 'a tradition'. They probably would not have wanted to be lumped together with some of the others I have linked them with. Moreover, they might well have argued, each one of them, that they had much else to say besides the above. Furthermore, various other traditions would like to lay claim to the work of a number of these. Positivists, for example, often look to Machiavelli, Voltaire, Hume, Darwin, Ferguson, Weber and Morgenthau as being hard headed (empirical) scientists who set out to take ethics and morality out of the study of human behaviour, especially political behaviour. Some postmodernists would like to claim Nietzsche or Foucault as guiding lights, if not founding fathers.

There is certainly truth in all this. But I think that the points of commonality I have made above still stand. I have not put these individuals together as an example of an emerging paradigm; whether they would have liked each other or could have worked together is neither here nor there. What is important is that in terms of the key elements of Enlightenment thinking, they all had similar doubts. And I think that their doubts led all of them to the general conclusions outlined above. Yes, realists were positivists in so far as they believed in science and the use of human reason; but they were much less sure about the existence of a rational universe with fixed laws waiting to be discovered.

Realists were postmodernists in that they wanted to blur distinctions between nature and nurture, human and animal, reason and fantasy, reality and illusion, but they were also scientists in the positivistic sense discussed above, and, I suspect, would not have been well disposed to postmodernists (on postmodernists, see below). Whatever else they were, realists were students of humanity. They believed it existed; they wanted to understand it, they probably even liked it; but they were not willing to unjustly glorify it. Although they did not believe in progress as a utopian end point or as a driving force of history, they generally believed that their own work could make life better for humanity. This was so even when there was a degree of despair. For example, as Freud famously suggested, 'the aim of the science of psychoanalysis is to help the totally desperate to be capable of sharing the ordinary miseries of life'. Or, if Voltaire's advice to 'tend our garden' turns out to be the best we can do, this might not be unworthy knowledge.

Whatever the case, their, intended or unintended, jolting of the classical tradition with their emphasis on a number of anti-enlightenment concepts, has been considerable. It is clear, however, that they have not created a 'paradigm shift'; what is less clear is, 'have they provided a sufficient basis for one?'. Certainly, cross-fertilization of ideas from these thinkers with the classical social sciences has been minimal. In general, in fact, the classical tradition has tried to ignore these theorists, rejected them as being irrelevant, sometimes holding them up as contemptible and dangerous, if not evil (see note one, p., 234). This was undoubtedly largely because of the realists relatively clear rejection of the notions of rationality, design, purpose, direction and progress in human social affairs - all dear to the heart of not only classicalists, but to numerous philosophers and theologians past and present. As noted, most of the propositions of the realists were, in fact, an attack on some of the basic notions of the Enlightenment itself. Moreover, a great deal of classical theorizing, intentionally or unintentionally, tended to *generate* oppositions between: rationality and emotionality, truth and falsity, natural and social, animal and human, biology and mind and, very significantly, between good and evil, whereas realists were working to obscure such distinctions.

Furthermore, many philosophers and classical social scientists had more or less eliminated power and politics as important theoretical concepts on the grounds that they represented selfishness, oppression, exploitation, violence, corruption, inefficiency, social instability and general irrationality. For most of the classicalists, power and politics as master concepts were direct enemies of grand abstractions such as 'society', 'capitalism', 'social systems', 'progress' and so forth. Also contributing to the rejection of realist thinkers was the fact that, despite strongly believing in reason, science and theorizing, the realists did not seem to produce work which easily lent itself to either positivistic empirical studies or to the development of large scale social theory. Studying history (sometimes speculatively), dreams, fantasies, literature or power politics did not especially impress logical positivists, the builders of functionalist grand theories or Marxists whose prominence began to be felt in the early to mid part of the Twentieth Century.

And in this the classical thinkers had, perhaps, a valid point. Realists did not seem able to translate human nature or the will to power, for example, into a theory of human 'society'. Their numerous observations and propositions concerning human passions and practical politics were not enough for those who wished to discover the 'laws', to be equal to those being developed in physics, to be used in theorizing the 'systematic workings' of human 'societies'. One of the most significant contributions of the classical social science tradition, one could argue, was its goal of attempting to develop a *social science* in which a theory of *social behaviour* was a major part rather than to subsume human social behaviour under classical biological, psychological, economic, liberal (rational) or historical theorizing. After all, humans are overpoweringly a social species. Social theorizing also aimed to dethrone self-interest from the pride of place it had gained and to overcome a perceived tendency to ignore moral issues in most areas of theorizing. Because, above all else, those in the classical tradition considered themselves to be scientists, social reformers and moralists, all wrapped up into one.

They realized that this would not be easy and that a lot of hard work would be necessary. With great enthusiasm they set out, as in all science, to discover empirical facts, which were meant to interact with social theory in order to produce ever more questions for empirical investigation. These would then contribute to theory refinement, and so it would go on. There were often arguments among classicalists between empiricists and theorists about the value of each others' work, but this, it was felt, was healthy, only fueling the scientific enterprise. Empiricists, for example, not yet willing to completely abandon the positivism of Comte, often had little time for grand theorizing, generally criticising the circular, teleological, and sometimes extremely abstract nature of much of it. In support of their position empiricists provided a considerable amount of knowledge about what came to be known as: 'cultures, economies, organizations, social institutions, political systems and social systems'.

On the other hand, there were those who argued that "Without a theory we are left swimming in an ocean of possibilities" (Hopkins, 1968, p., 71). That is, theorists maintained, without a theory we do not know what significance our facts might have in explaining social stability or social change, and without theory we would not know what further information was useful, what studies to carry out, what questions to ask. Moreover, they argued, 'facts' can, and should, be threatened by theories - something in effect realists would undoubtedly have agreed with. Taken for granted facts, facts which are primarily functions of particular methodologies, politically inspired facts, for example, can be very effectively shown to be in need of correction, or even in need of being discarded, when demonstrated to be theoretically untenable. And the more inter-disciplinary based the theory, the more value it is in scrutinizing facts derived from narrow perspectives - that is, in overcoming the intellectual resistance of 'experts' who have a tendency to ". . . never threaten a fact with a theory" (Harris, 1977, p., 8).

Thus classical theorists were busy establishing the science that realists so admired but themselves seemed so unable to achieve. As noted, realists appeared to have left a void (as had metaphysicians before them) in that they could not jump from micro concerns, or one or two theoretical issues (for example, reproduction, self-images, power/politics), to a comprehensive theory of 'nature' or 'society' which had predictive, reforming and morality improving powers. On the other hand, it can be argued that the concern of leading members of the classical tradition to develop an over-arching theory of 'society', with morality-improving implications, meant that they became embroiled in the circularities and teleologies of most of modern social theory. For classicalists, cultural, institutional, social, structural and market facts, facts about 'patterns of culture', the 'division of labour', 'supply and demand', about class structures and about human social systems, amounted to theories about the systematic nature of human societies and how they worked (functioned) in an abstract, perfectible sense.

What may well be the case is that we should not seek a theory of 'society' as such; instead we should look to theoretical linkages of human motivations, social interactions and power politics (from everyday expressions of a human will to power to the politics of the state) with a view to explaining the *apparent* stability and organizational character of human social life without assuming that it has structural, systematic or spiritual qualities. We require a theory of the *processes* of human motivations and social interactions, not a theory of human perfectibility, social stability or inevitable progress, because, worthy as these concerns are, they too easily detract us from a realistic theory of human behaviour. Our theory may not end up being a universal theory of everything with the capacity to predict the future, but then neither is Darwinian theory, irrefutably one of the most useful foundations for science during the last 150 or so years.

Probably the best we can hope for, then, is a theoretical approach which allows us to ask the right questions and make generalizations about human motivations, social interactions and power politics. (If a theory of justice can benefit from this, all to the good, but it must not be an overriding concern at this point.) I will argue that this is indeed the case, and that the realists give us a start. And when we add contributions from recent Darwinian challenges to enlightenment thinking, such as from sociobiology and psychobiology, the prospects are moved forward. But first we must give the classical tradition every chance because, while struggling for a theory of society, they came to realize many of its teleological and circular defects and set out to overcome them.

Is There Life Left In The Enlightenment Social Sciences?: Critical Theory

By the early part of the Twentieth Century, for example, certain philosophers and social scientists brought up in the classical tradition (especially the Marxist and Weberian traditions, and later some of those concerned with feminist issues) began to recognize a number of its drawbacks, and also to see some value in the work of at least some of the members of what I am calling the realist tradition. They set out to make amends for what they considered to be the naiveté of many of the Enlightenment assumptions found in the classical tradition and to bring skepticism, sexuality, self and power, among the conceptual mainstays of the realist tradition, into social theorizing generally, without losing, they hoped, the promise of reason and progress.

Critical theory (ranging from aspects of Marx's work to that of the Frankfurt School of the early to mid 1900s - Cf., Fromm, 1960; Marcuse, 1962, 1964; Adorno, *et. al,* 1969; Bauman, 1976; Horkheimer and Adorno, 1972), for example, challenged the notion of 'instrumental rationality' as being the major driving force behind some vague inevitable evolution to utopia. Indeed, instrumental rationality was seen as a major means of maintaining *domination* by certain others and 'the system' over the majority of people. It did so by turning individuals into *objects*. 'One-dimensional culture', that is, instrumental culture, became the umbrella oppressor. 'False needs' were the opiates through which it addicted almost everyone - the satisfaction of which provided 'false security'; eventually extremely dependent 'false selves' emerged. These were, it was claimed, easily subject to political domination by charismatic 'satisfiers' of false needs, extremist political parties and leaders.

Critical theorists had little in the way of advice for escape (Craib, 1984), but they did propose *critique* of existing social arrangements and advocated somehow taking control of self, social relationships and social evolution. The study of language, in a search for meanings related to issues of control and free will, and to allow for a de-construction and then re-construction of existing social forms, was often suggested as a means to this end. Marginal groups (for example, students, blacks) were often seen (instead of workers) to be the potential spearhead of revolution. Unfortunately, despite providing useful insights into human behaviour and valuable criticisms of the traditional social sciences, certain social practices and many dimensions of modern societies, a number of un-substantiated assumptions and postulations (often ones that would be difficult to substantiate at the best of times) were made by critical theorists. These were most often, implicitly, if not explicitly, part of a process of attempting to identify and rid the world of 'evil' (often thought of as capitalist repression). As such these attempts resulted in a considerable amount of rather ill defined, (sometimes mystical) notions of 'good' and 'evil' (for example, freedom/equality vs. oppression/exploitation, 'true-self' vs. false self/consciousness, 'authentic' vs. 'in-authentic-self', 'authentic' vs 'in-authentic' relationships', and so on). Critical theorists have, as a result, been accused of being elitist because often *their* (meaning intellectual professors') perceptions of such things as 'true-self' and 'authentic relationships' were being claimed as representing universal 'truths' and 'the good' (Craib, 1984).

In more recent times, feminist theory too has taken a very critical stance with regard to both traditional social theorizing and modern social life (Cf., Keohane, 1982; Hutcheon, 1989; Lovibond, 1989; Lengermann and Niebrugge-Brantley, 1992). Feminist theory held the promise of re-examining many of the rationalistic assumptions of the Enlightenment, of bringing emotions (passions) back into theorizing and of attempting to link human nature to personal relationships and these to social behaviour. In the event, however, this has not been the agenda: ". . .feminist theory is critical and activist on behalf of women, seeking to produce a better world for women - and thus, it argues, for humankind" (Lengermann and Niebrugge-Brantley, 1992, p., 447). This is fine, but what does it mean in terms of the realist concerns, especially the evolutionary and psychological ones? Will they play much part in feminist scholarship; or will the desires about what society *should* be like, based on Enlightenment optimism, overly influence it?

Feminist theory, in fact, has generally set out to distance human behaviour from biology - especially sexuality based on biology (but see Rossi, 1977; Hrdy, 1981; Zihlman, 1981; Fedigan, 1986). It has also set out to move away from psychology and the apparently 'non-rational', and even 'immoral', male enthusiasm for sex, competition, war and power politics - all concerns which were/are basic to the realists. Above all, most of feminist theory has been devoted to 'demonstrating' that there are no fundamental differences between males and females; those that appear to exist are attributed to social causes in which maleness has somehow gone wrong. Such sociological approaches move feminist theory a long way from realist concerns in favour of systems explanations. Therefore, the same criticisms and problems of theory discussed in terms of the traditional social sciences and critical theory generally also apply to them.

Even 'psychoanalytic feminist theory' (Cf., Chodorow, 1978, 1990; Dinnerstein, 1976) does not really probe bio-psychology but rather looks to infant and early childhood experiences as causes. Moreover,

> "Psychoanalytical feminists operate with a particular model of patriarchy. Like all oppression theorists, they see patriarchy as a system in which men subjugate women, a universal system, pervasive in its social organization, durable over time and space, and triumphantly maintained in the face of occasional challenges. Distinctive to psychoanalytic feminists, however, is the view that this system is one that all men, in their individual daily actions, work continuously and energetically to create and sustain" (Lengermann and Niebrugge-Brantley, 1992, p., 471).

Nevertheless, such theory does at least try to incorporate some of the irrational, non-instrumental dimensions of human existence into its explanations. Moreover, psychoanalytic theory highlights the one area of 'system' or 'structure' which feminist theorists can claim as their own; that is, the notion of patriarchy. Is there something in human nature, derived from evolution, which results in the emotions and power relationships which have been defined as patriarchy? And, given the centrality of reproduction in the evolutionary paradigm, do these power relationships pre-date, and superordinate, other social relations? If the answer is yes, feminist theorists have a basis which is equal to, if not greater than, Marxists, socialists, functionalists, liberals or psychoanalytical theorists in claiming that their theory should be the mother of theories.

Given the philosophical stance of feminism noted earlier, however, it is not likely that feminist theory will wish to follow the above possibility. Certainly, to this point it has not. Generally, patriarchy has been defined as 'men hate and dominate women' or as a *system* which provides for the continual exploitation and oppression of women, the 'fact' of which is treated as sufficient *cause* of patriarchy itself. Radical feminists, for example, have made oppression the central plank of their approach. For them it is everywhere, all pervasive; the only possibility of escape is through continual confrontation and/or separation of the sexes.

Feminist theory has gone a long way to bring women into the theoretical equation; it has appropriately emphasized the importance of considering power relationships in social theory; feminist theory has been instrumental in making us aware that 'the private is political'; it has insisted that individual feelings count; it has highlighted a number of areas in human social life which have been unfair and it has shown how much of the science which has gone on before has been distorted by male bias and by a number of Enlightenment assumptions about a completely rational universe. It has not, however, really challenged the Enlightenment, nor has it challenged classical social theory. The philosophical issues of teleology, holistic determinism, final cause, purpose, meaning, directionality, human nature, a will to power, the role of emotions in human social life and power politics have been largely ignored in the quest for discovering social causation and the importance of fighting *oppression.*

Critical theorists, including the feminist variety, then, have not escaped most of the spiritual assumptions, circularities and teleologies they had often criticized when found in the classicalist's premise of an inevitable human progression through the use of instrumental rationality to some sort of perfect state. The details of what caused good or evil changed somewhat, but the spiritual quest for the 'good' remained. Furthermore, their own assumptions and postulations about the possibilities and characteristics of social/moral progress (towards non-alienating, non-repressive, non-exploitative life conditions) have not been any more established by empirical 'fact' or theoretical scrutiny, than those of the classical tradition they rejected. Critical and feminist theory, then, have raised important questions, and have addressed, and affected, numerous issues in which injustices clearly seem to have existed, but have done little in the way of advancing us towards an objective understanding of fundamental human motivations/behaviour.

Are We In Danger Of Throwing Out The Baby And The Bathwater?: Postmodernism

More recently, in dismay perhaps, postmodernist theorists too have taken on board many of the propositions of the 'realists'. That is, they have criticized the traditional social sciences' reliance on instrumental rationality, teleology generally, and many of the moral assumptions which derived from them. And they have gone further than this. They have also denied the whole notion of progress (indeed, the idea of society as an above human essence and the possibility of a ‘greater’ purpose). They have put forward the view that distinctions between fact and myth, between reality and images, truth and falsity, order and chaos and, more generally, good and evil, for example - so common in the Enlightenment sciences, including most of critical theory - are false distinctions. All we can know about, they tend to argue, are our methods of understanding, our methods of creating meanings (and thus cultural norms and values), our uses of language, our processes of interacting. As with the realists, (and interestingly enough, Hegel) everything is becoming, there is no being (For examples and/or discussions of this approach see: Baudrillard, 1983; Lyotard, J., 1984; Ross, 1988; Lovibond, 1989; Seidman and Wagner, 1990; Lash, 1990; Best and Kellner, 1991).

These ideas are attractive for those who fear the use of 'instrumental rationality' and utopian thinking in, for example, genocide and war (Cf., Bauman, 1988, 1989; Morgenthau, 1946; Halberstam, 1969). Moreover, the writings of postmodernists are both interesting and enlightening for those who are dubious about the notion of progress, who want to get rid of boundaries between disciplines, and who tend to see human behaviour and social life as processes rather than as 'things' or structures (Cf ., Hatab, 1995). Nevertheless, because of their total relativity, "radical relativism" (D'Amico, 1986), postmodernists do not take us very far along the way towards an understanding of the human condition. Indeed, while sharing many ideas and concepts with the realists, postmodernists generally disdain such notions as causes, science, and even theory, all basic to the realists. There certainly is little attempt to establish human nature as a basis for understanding human behaviour (but see Haraway, 1989), which was another key to the work of the realists (and some critical theorists - Cf., Fromm, 1960). As a result of all this relativity and 'rejecting', postmodernism does not make up a coherent body of work let alone a unified theory (Ritzer, 1992; Best and Kellner, 1991; Berman, 1982). In the words of one critic, admittedly one not well disposed to them, it is even worse than this: postmodernists offer ". . .definitions which [are] mutually inconsistent, internally contradictory and/ or hopelessly vague"(Callinicos, 1990. p., 2, see also, 1985).

In one sense this does not really matter. As long as they have interesting things to say, why should they be consistent and always crystal clear? No reason at all. However, the term theory should then be dropped. Theory means seeking linkages, causes, internal consistency, asking questions that most often require empirical investigation, more theorizing and so on. In other words, theory is a major part of science, and the realists were very interested in science. Postmodernism clearly is not a science; postmodernists most often do not want it to be a science. Those few who have not rejected science completely are at best interested in ". . . a break with Newtonian determinism, Cartesian dualism, and representational epistemology. . . [they]. . . embrace principles of chaos, indeterminacy, and hermeneutics, with some calling for a 're-enchantment of nature'" (Best and Kellner, 1991, p 28). It seems that postmodernists do not want to expose their thinking to either facts or theories, since they have little faith in either.

IN SEARCH OF AN EMPIRICAL SCIENCE OF HUMAN BEHAVIOUR

So what are we to do? Do we accept the view that everything is so relative that we cannot know or make claims about anything? Is knowledge concerning the human condition beyond us? Put another way, is the science of the Enlightenment - the science which has done so well in understanding physical nature – inappropriate when it comes to understanding human thoughts, behaviour and history? In answer I would say that science in the form of social science may have been disappointing so far, but I do not think it has completely failed us, nor do I think that it is inappropriate for an understanding of human behaviour. If we take the view that science is a means of understanding *everything,* and for developing *laws* about human thoughts, behaviour and history equal to the laws of Newtonian physics, then, yes, social science has failed us and will continue to do so, and probably does not deserve the title 'science'. But if by science we mean systematically gathering information (continually looking for new and relevant information) and endlessly conceptualizing and theorizing about *causes* in a non-circular, non-teleological manner, social science is not only viable but can lead us to understand a great deal about the human condition.

Cause is important here (see Chapter Three) because it is precisely due to the fact that the assumptions, methods and theorizing of the traditional social sciences, critical theorists and postmodernists, have not, for the most part, given us any clear notion of *universal, non-teleological causes*, that we are scarcely better able to understand basic human motivations than any socially perceptive person; certainly less able than most successful lovers and almost all successful politicians. Unfortunately, certain critical theorists and all postmodernists have largely rejected the notion of cause altogether, in some cases treating those who search for causes as being, at best, 'old fashioned' and misguided and, at worst, politically harmful. If, however, we cannot anticipate behaviour, social conditions or events from their precedents, if we cannot know what makes us *want* to do certain things, we have no hope of *understanding* human behaviour. A study of causes is our chance to escape the problem of total relativity but they must be non-teleologically derived causes. In many ways a *search* for causes, in this sense, gives the concept of social science meaning.

Following this view, an attempt will be made to outline a possible basis for a social science which relies on at least some of the non-teleologically derived, non-utopian principles of nature first uncovered by physical and biological scientists. This will start by considering how we might be able to determine what it is to be human in a fundamental sense, to go some way towards giving the social sciences what 'the laws of gravity' and 'reproductive consistency' (later known as genetic replication) gave the physical and biological sciences; that is, a *consistency* against which to assess, and sometimes predict, *variation*.

I will be maintaining that if we can establish that humans are 'such and such' we will be in a position to suggest what might be congenial, or conversely, in conflict with that 'such and such'. We would, for example, be able to ascertain, and possibly predict, the conditions in which humans might desire or fear the exercising of power/control - by themselves or by others over themselves. We could assess the extent to which different patterns and degrees of hierarchical relationships might be congenial or in violation of the 'such and such' of humanness; the conditions under which humans are likely to feel deprived, happy, successful or failures; and from this the conditions under which humans want to subordinate or revolt . We may reveal the extent to which certain types of child rearing and early experiences leave an individual vulnerable to the above.

History (along with studies from anthropology, psychology, economics, sociology and political science) will have information from past and present events which give us an indication of the safety and/or dangerousness perceived by humans faced with various events, social patterns and institutions, helping us confirm, or not, our theoretical speculations concerning the such and such of human nature, while at the same time suggesting reoccurring patterns of behaviour. We should be able to ascertain the degree of fragility (Turner, 1993), and thus potential changeability, of at least some of the 'institutions' of civil (political) society. We would be working towards substantiating Fromm's (1960) claim that "What is good or bad for man is not a metaphysical question, but an empirical one that can be answered on the basis of an analysis of man's nature and the effect which certain conditions have on him" (p., 229-230).

So far all we have are: the (often wishful) spiritual and teleological value judgments of the traditional social sciences; a rejection of the realist attempt at a causal science of humans (largely because of the realists' rejection of a number of the value assumptions common in the classical tradition); the sometimes intellectually elitist, sometimes moral, preachings of critical theorists (and some feminist theorists); and the total relativity of postmodernists. We also have a lack of successful theorizing on the part of all these, including the realists, when it comes to human social 'organization' for the purpose of administering social life. Perhaps this is all we can hope for. It might be that postmodernists are right; maybe all knowledge (human conceived truth) is totally relative; maybe postmodernists win. But this leaves us with the very disturbing implication: can Nazi, feminist, capitalist, socialist, pro-abortionist, 'right to life-ers' and so on, all be right, all have equally valid claims concerning the search for knowledge, 'truth' and justice? We need not, I think, concede all this yet.

In order to go beyond the trap of total relativity, however, we not only have to discover the essence of being human non-spiritually and non-teleologically, but also independently of our own wishes, and independently of what our culture claims it is or should be. That is, if we are to sound more convincing than moral or ideological claimants, we must divorce our own concepts of what we think humans, or human societies, need in order to fulfil the purposes we have assigned to them, from our analysis of what humans really are. In the task of outlining such an approach. I will be arguing that most of the suspicions of the realists concerning the 'theologies', teleologies and circularities of the classical tradition are valid. Also it will be argued that we should accept much of the realists' extremely valuable intellectual contribution to our understanding of human consciousness, motivations of behaviour and practical politics (political society). At the same time, we should not ignore some of the substantial findings and thinking which have come from the classical tradition (among whose members, each in their own way, there often was a recognition of certain drawbacks of their paradigm, as well as acknowledgment of the value to be found in some of the ideas from what I am calling the realist tradition).

The theoretical work undertaken here will specifically wish to avoid the grand circularities and teleologies of much of the corporatist and functionalist traditions during their grand theorizing and systems-building periods. In their place, it will be argued, a concept of ‘political society’ must be fashioned, one which is *linkable* to human nature. The notion of political society suggests that what *appears* to some to be moral wholes or ‘social systems’, are really products of *power networks* (patterns of: camaraderie, alliances, patron/client relations, domination, subordination, hierarchy, for example) constantly generated and maintained through *practical politics*.

Fortunately, recent work in the biological and physical sciences has made it possible to begin to develop non-teleological concepts of the bio-electric dimensions of human nature which are consistent with known laws of physics. Social scientists, I will be arguing, can complement these developments by searching for universals of: human emotions, human desires, senses of danger, cognitive patterns, patterns of social behaviour and patterns of power politics through comparative (cross-cultural and historical) studies.

However, if findings from brain sciences and universals discovered through comparative studies are to be put together into one science, conceptual mechanisms which link them must also be developed. This will depend upon theoretically linking human genetic effects, physiological processses, psychological reactions, social behaviour, politics and cultural creation into one scheme. Unfortunately, in the past such attempts have generated a considerable amount of anxiety and antagonism. This is undoubtedly because at some point on this continuum, usually between physiology and psychology or between biologically orientated psychology and social psychology, a major separation has been made in Western thought; that is, the separation between mind and body; man and animal; rationality and irrationality; a separation which if breached means that matter and soul must uneasily meet, that 'animal passions' will threaten spirituality, that 'rationality' will be forced to face the onslaught of 'irrationality'. It is in this merging that the Devil threatens God and it is where the human 'purpose' towards utopia might be side-tracked.

But it is also here that biology becomes self and as such individuals become social actors carrying the biological traditions - the consistency that is the species *Homo sapiens* - into the future. Therefore, despite the anxieties and objections, we must seek a scientific understanding of this linkage; this will be the major task of this work. However, before directly addressing the characteristics of human nature (the bio-psychological processes which link physiology and self), the next two chapters will focus specifically on the 'moral' and scientific issues involved in breaching the gap between biology and behaviour, because until these are resolved there is little hope of the long term development of a science of human behaviour.

SUMMARY AND CONCLUSIONS

This book will seek to develop a model of *bio-psychological* human nature which is linkable to a notion of *politics* and *political society*. The aim will be to work towards the development of a non-teleological social science which is not in conflict with Darwinian principles of evolution or with the principles of genetics and bio-electricity current in the physics of brain science, and at the same time utilizes both empirical evidence and theoretical propositions derived from the social sciences.

I will not always systematically go through other possible explanations to the one being forwarded regarding specific issues. This may be unfortunate; however, it is not the major purpose of this work to analyze theory or theoretical contentions or theoretical controversies. Rather, the purpose is to outline an approach which might solve a number of problems currently extant in the social sciences and to try to incorporate current work from evolutionary theory into our social science. The search is not for a 'theory of everything' but instead we are looking for some new grounds for *theorizing* so that new questions can be asked, new concepts developed and new understandings achieved. The aim is to do this avoiding as many as possible of the very questionable Enlightenment assumptions (especially those about rationality, perfectibility and progress) which have plagued modern philosophy and the social sciences. To have looked at every possible interpretation of some of the arguments which will be presented would have greatly distracted from the main task and, anyway, would have made this work much too long. I hope that the story tells itself in such a way that readers will see other possible interpretations when appropriate and be able to weigh up various developments and then move on to the next step in the general argument.

Specifically, I will be looking at material from: the neural sciences, various schools of psychology, paleoanthropology (human evolution), sexology and the social sciences. Speculation will be involved. Some things will be postulated which have not yet been fully substantiated by the test of hard empirical evidence. At the same time, however, current work in all the above areas will be used and the aim will be to claim nothing which has been refuted by empirical evidence. The goal is to follow the notion of "consilience", where scientific integration is the major task and in which the unifying principle throws light on sub-areas and these combine to give credence to the unifying principle (Ruse, 1986). It is intended to be a scientific process in which observational evidence will be used to generate theory, which will lead us to other questions, a search for more evidence, refinement of theory, and so it should go on. In this spirit, it is time now to consider how biology and social science might work together in order to begin to determine how biology becomes self.

CHAPTER TWO
BIOLOGY AND THE SOCIAL SCIENCES: THE ISSUE OF MORALITY

A first step in creating a science of human behaviour will be to link conceptually human genetic effects, physiological processes and psychological reactions into a model of human nature. If this model is to be non-teleologically derived, and non-spiritual in characteristics, biologists and social scientists will have to work together to ascertain what is really there in terms of observations rather than what we wish to be there, or postulate *should* be there. Moreover, we require much more than a meeting of 'representatives' from both fields with a remit to put together what is already known in each field, or a division of labour in order to uncover additional information from two distinct fields. The fields of study must themselves be *conceptually merged* if the barrier between them is to be surmounted. To do so, we require new concepts, conceptual frameworks and, indeed, theoretical orientations, which contain fusion within themselves. This requires joint conceptual development.

This raises problems, however, because it means overcoming the psychological barrier - often perceived as a spiritual barrier - which has been erected between variously: animal and human, physical and social, instincts and reason, bestiality and morality. Unfortunately, in no small part, social scientists and biologists and, especially social scientists, have *themselves* created the barrier during the last one hundred-plus years since Darwin first published his *Origin of Species* (1859). This has been done, for example, through the 'elevation' and declared independence from biology of 'rational man', through a claim for cultural supremacy and/or postulations of the causal and moral sovereignty of society/social systems. This includes many of those who were specifically looking to understand human consciousness (for example, behavioural and social psychologists). Such psychologists were more interested in bridging the gap between mind and society than between brain and behaviour. They were, in fact, particularly insistent on distancing psychology from biology and allying it with sociology. In the process, mind became a product of society with any possible biological connection greatly downplayed, if not forgotten or steadfastly denied.

With the arrival of sociobiology in the late 1960s and 1970s, however, this separation came under challenge. While some early sociobiologists - followed by psychobiologists - seemed to want to absorb, or, indeed, eliminate, the social sciences (Cf., Wilson, 1975; Symons, 1979; Barkow, Cosmides and Tooby, 1992), certain philosophers (Cf., Ruse, 1979, 1986; Dennett, 1995; Rosenberg, 1981; Trigg, 1982) looked for a fusion of the two disciplinary areas. At the same time, certain biologists and psychologists -

and occasionally other social scientists - began to made attempts at formulating linkages between them. Nevertheless, very little in the way of fusion concepts have been developed. We still do not have a unified bio-psy-chological approach which offers a seamless transition from genes to neuro-biology to psychology to behaviour to society.

For example, whatever the psychobiology or sociobiology on offer, (see next chapter), there has been a prevailing tendency to jump directly from principles of sexual selection to *abstract categories* of behaviour - such as kinship, altruism and reciprocity - thus, avoiding the gap between genes and behaviour. At the same time, a rationalistic cost-benefit analysis of the effects of selfish genes seems to indicate the evolution of 'rational man', very capable of operating co-operatively with kin and in patterns of reciprocal altruism (but, still being able to compete effectively with rivals). Very recently some scholars *have* put forward an argument for the evolution of a Machiavellian intelligence (Cf., Byrne and Whiten, 1988; Whiten and Byrne, 1997; Dunbar, 1993, 1996) as the link between genes and behaviour. Others have advocated mind-as-computer (or at least as a computational system) as the linking process (Cf., Cosmides, *et. al*, 1992; Tooby and Cosmides, 1992; Pinker, 1998). In both cases, in a circular fashion, these approaches continue to rely on the social categories of kinship, altruism, co-operation and reciprocity, as the *causes/descriptions* of the calculating Machiavellian intelligence and/or operations of mind-as-computer.

Sympathetic social scientists and some philosophers - unwittingly following Spencer - often simply use the ideas of natural selection as an *analogy* to explain *why* existing ideas or patterns of behaviour in human social life have come to be - because they were *selected*, therefore, presumably, they were/are the *best*, is the view. In these approaches, biology and behaviour, quite apart from genes and behaviour, continue to operate at quiet distinct levels (Cf.., Parsons, 1966; Richardson and Boyd, 1978; Granovetter, 1979; Irons, 1979; Dennett, 1995; Runciman, 1998, 1998a). During all of this, despite both areas desperately avoiding the gap, the anxiety and antagonism between biologists and social scientists has not abated. Indeed, it often seems to have become greater with time (Cf., for social scientists - Allen, 1975; Sahlins, 1976, 1996; Lewontin, *et. al*, 1984; for biologists - Barkow, Cosmides and Tooby, 1992; Symons, 1992; Pinker, 1998), especially, it appears, with the increasing success of the Darwinian 'paradigm' in exploring and explaining primate, including human, behaviour (for a discussion see: Irons and Cronk, 2000; Gray, 2000).

Why might this be so? How might we account for the persistence of the split (the tendency to jump the gap between biology and human behaviour rather than to conceptually fuse it, even among those who argue that a gap should not exist) and why the continuing antagonism between two scientific areas? These are important questions because answers should help us understand some of the ideological, psychological, spiritual and moral difficulties which stand in the way of conceptually uniting biology and the social sciences. Answers should also help us understand more clearly what it is we have to do to dissolve the barrier between the two areas, if it is, in fact, possible to do so. We can usefully begin our search for a solution to this problem by pondering the full intensity, staying power and significance of the *emotional* impact of Darwin's contribution on our perception of human nature and the human condition generally.

DARWIN'S IMPACT

Arguably, Darwin has had the greatest impact on human consciousness since Copernicus, perhaps since the founders of the great world religions (see also, Desmond and Moore, 1991; Dennett, 1995; Gould, 1996a; Jones, 1999). Once Copernicus was accepted, however, that was the end of the matter. This has not been the case with Darwin. His ideas fuelled controversy at the time and ever since; scientists, philosophers, theologians, politicians, journalists and business people have all entered into the debates.

From the beginning there were those who were ready to believe and to defend and preach his message with an almost religious fervour. This included some of the most famous scientific minds of his day (but not all by any means). Darwin also had the support of some of the most respected 'gentlemen' and leaders of his time. His contribution was considered to be of such importance that he is buried in Westminster Abby. It was not all positive, however. There were fierce critics, passionate denunciations from scientists, religious leaders, social reformers and political notables. More than half a century later a famous show trial was held in the USA to attempt to counter laws which made teaching his ideas illegal - with most of the world's media and many of its scientists in attendance. This issue is still not settled; in parts of the USA the legal battle rages on. And in the sciences, the relatively recent arrival of sociobiology has regenerated the debates originally initiated by Darwin. The capacity for a *scientific formulation* to continuously maintain such intellectual and spiritual unrest, even fear, seems unequalled; it is, as noted, almost in a class with world religions.

In one sense, all this fuss seems quite amazing given that Darwin was by all accounts a mild mannered man who did not especially set out to upset most of his own and every subsequent generation. Darwin preached no specific religious, social or political ideas; he was far from revolutionary; indeed, he was a relatively conservative individual. There have been no spiritual or political movements or political parties based on his ideas; no political state has been formed around his precepts.

So what was the message that caused, and continues to cause, such controversy? The first point of significance is that Darwin's inquiry into the geological history of the Earth and the *processes* of the *origin* of species upon it, pierced deeply into the very soul of humankind. It raised the issue of 'who are we?', 'why are we here?', 'where are we going?' and, probably most importantly of all, 'is there any reason for all this'? (Darwin also, of course, raised the issues of science vs. religion and modernity vs. tradition.) But so what? After all, these questions had been raised by numerous theologians and philosophers time after time. The difference undoubtedly is that the answers before Darwin usually came out quite 'positive' – the perfect community or polis, or failing that, a heaven in the hereafter, were possible, indeed achievable. In the cases where more 'negative' thinking emerged, it died away relatively quickly, or was confined to obscure religious or philosophical cults, or else attributed to deviant thinking (such as in the thinking of the Sophists, Sceptics, Manicheans, Epicureans, Lucretius, Marcus Aurelius, Machiavelli, Hobbes and Malthus, for example).

On the face of it, Darwin's ideas were quite positive. It was possible to interpret his contribution to mean: things in nature change, the old order is *not* fixed by God or set forever by tradition, evolution is inevitable, social life is a competition in which merit (the best) must inevitably win and, therefore, progress is not only possible but part of the very nature of the universe. And science, not tradition, theology or metaphysics, is the way to understand the universe. These were, for many, the lessons of Darwin (Appleman, 1979; Desmond, 1989; Desmond and Moore, 1991; Bowler, 1989; Eiseley, 1958; Gay, 1993; papers in Maasen, Mendelsohn and Weingart, 1995). On careful scrutiny, however, Darwin's contribution contained an altogether different message; it was not a very congenial message. To some it came across that God was dead (Nietzsche's phrase), that nature was 'red in tooth and claw' (the poet Tennyson's phrase), that the world was an arena where 'survival of the fittest' (Herbert Spencer's phrase) and 'dog-eat-dog' (Jack London's phrase) ruled - it certainly was no place for the weak - and, to add insult to injury, we were part monkey.

This is poetic and philosophical licence bearing little understanding of the nature of Darwin's concept of fitness let alone of natural or sexual selection. There is, however, no doubt that Darwin's concepts of natural and sexual selection took teleology, direction, purpose and progress out of geology and biology, including human biology. This was often hard to take, even among many of his scientific colleagues. Nevertheless, as we have seen, there were a considerable number of very important people, scientific and otherwise, ready to believe in and defend Darwin with great intensity - some who understood the full impact of his ideas and some who did not. So, from the very beginning, Darwin's message was ambivalent, confusing. Its impact was not, and has never been, directly obvious. It could be interpreted in a variety of ways, often greatly influenced by the social, religious and political positions of his admirers and critics. Although he was attacked when he first put forward his 'revolutionary' ideas of natural and sexual selection, it was not with the same intensity of that unleashed against, say, Bruno or Galileo - at least not officially. He was held up as a devil figure by many, but not to the same extent as Machiavelli, Freud or Marx have been at various times. There was never any official hint that he should be tried as a heretic or lose his citizenship rights; indeed, as noted, he was buried in Westminster Abbey. (But, on the other hand, he was never knighted).

Moreover, although the scientific significance of Darwin's view was recognized at once, at least by scientists, it was somewhat ambivalently received; it was not perceived with the same dramatic sense of its potential value as the work of, for example, Newton or Einstein. Nevertheless, in the scientific world, rejecting Darwinism has come to be a bit like being in favour of the jailing of Galileo or the burning of Bruno or taking out membership of the 'Flat Earth Society'. Scientific people are almost universally Darwinians - as are most non-scientists who like to consider themselves educated and modern. Darwin more than any other figure has come to represent modernity and science in the face of tradition, religion and superstition in the modern world. Darwinism, despite its anxiety-generating aspects, takes on, in the minds of many, 'superstition' and tradition; it compels admiration as a pinnacle of modernity, of human science and rationality leading us away from backwardness; it remains as modernity winning out over religious dogma and non-reason, of human freedom being gained from the destruction of the notion that social hierarchies are God given, right and here for all time. It is science as opposed to theology and metaphysics.

So, despite early worries and rejections, with these accomplishments, with these accolades, with this historical significance, how is it possible that Darwinism *continues* to generate such unease (and often quite negative passions) in so many circles, including scientific circles? Put another way, given the scientific pedigree of Darwin, how has biology increasingly come to be seen as such an enemy of humanity? After all, after initial problems, science and religion have reached an accord and have been able to live together peacefully. Individuals either believe totally in one and reject the other or, more likely, mentally separate them and accept both, ignoring possible conflicts between them. Indeed, even most official religious bodies have come to terms with science. Why cannot the same be said for evolutionary biology and the social sciences? Why does Darwinism continue to rear its head as a threat in ways that science no longer does?

A large part of the answer must surely be that the seemingly non-rational, non-purposive, non-design approach to the key philosophical questions raised, cannot be completely ignored, rationalized or dismissed by humans. God may have been dethroned and science installed, but nature, uncontrollable, purposeless, directionless, unpredictable, *uncaring* seems to have taken his/her place. Genes rather than free-will and reason are, it appears, offered as causes of behaviour. The concepts of natural and sexual selection, when fully applied, leave little if any room for notions of rationality, design or progress. Most scientists, let alone humanists, theologians and the general public, find this difficult to accept.

But, significantly, as time has gone on, Darwin's paradigm shift has replaced almost all previous theories in paleo-geology and biology. In the process it has unified geology, paleontology, anatomy, bio-chemistry, physiology, genetics (not known about in Darwin's circles), animal behaviour, ethology, sexology and psychology into one framework. This framework has gone from strength to strength, replacing all other evolutionary paradigms (concerned with nature). It continues to thrive and increasingly seems impossible to marginalize. Palaeontological evidence (including that pertinent to human evolution), the discovery of the structure of DNA, the genome project, genetic engineering and advances in medicine all seem to support the value of this paradigm. Indeed, there are some who have come to believe that biological (especially genetic) understanding will provide the basis of human perfectibility. Yet, for many, all of this 'success' only suggests that science is perverting nature, that we are in danger of creating a deified science to take the place of a real search for self-actualization and morality and that, anyway, we are in for a big disappointment in our Frankenstein.

Darwin, therefore, was, and is, very difficult to *fully* accept or reject. His 'paradigm shift' *continues* to affect people in ambivalent, sometimes opposing, ways. Copernicus, Newton and Einstein (despite temporarily decentering humankind and creating the physics of an atomic bomb) were finally accepted and that was the end of it. In fact, when the dust settled, these could be considered discoverers of God's laws. Design, direction and a moral universe were not disproved by science, just the opposite, they were proved by science. Some of the realists listed with Darwin in the previous Chapter have also received a lot of press, much of it negative. But the vast majority of humans who know about them have been able to completely ignore them or, when this is not possible, to demonize them and move on. Hobbes, Machiavelli, Nietzsche and Freud, for example, have not fared well in terms of popularity in either scientific or popular circles. But they can easily be left out of explanations or treated as exemplars of 'bad thinking' and 'bad practice'. They can be scoffed at as examples of individuals obsessed with power, supermen and sex, to be joked about, to tease people with statements like 'that's Machiavellian', 'that's Freudian' and so on; these dissidents of the enlightenment are not to be taken too seriously.

Even when they are presented as serious enemies of humankind, it is in order to 'show how' their ideas can be overcome with right thinking and good practice. These 'devils' often served as intellectual scapegoats to highlight just what was, and is, considered to be good theory, practice and moral behaviour. Even the intense criticism of the Church by Machiavelli, and Christianity by Voltaire and Freud, gained them less opprobrium from most strongly committed Christians than Darwin who was quite ambivalent about, almost sympathetic to, Christianity. Marx, the most demonized of the Classical tradition, who specifically set out to develop 'a materialist conception' of history, generated only time and location specific controversy or denunciation. These are all easy to ignore (by lay people and scientists alike). Darwin's explanation and its implications will not go away; they seem to get stronger with every attack upon it.

Darwin's controversy is of long standing and still continues, then, in large part because it is not easy to accept, ignore or stigmatize his work; his impact was not direct, the danger was never clear, nor is it now. As noted, he had (and has) critics but he also had (and has) a devoted following. It is difficult to ignore - or reject - a 'paradigm' which seems to explain so much, which has unified so much and which has been progressively strengthened by continual mounting evidence. Yet it is equally difficult to accept that the universe is a place where 'random' chance has a significant part and where there is no grand design, purpose, direction, or progress. If this is the case can we have a notion of 'truth' or of 'morality'? How can these exist if the universe is purposeless and meaningless? Does the notion of '*modern*' have any relevance? These are the perplexing questions left unanswered. These are questions which will not go away.

So what to do? Biologists gradually 'solved' the 'dilemma of Darwinism' by setting out to tame natural and sexual selection. Selection became a process of 'adaptation'. Unfortunately this became more than Darwin's process of organisms more or less 'beautifully' adapting to changing circumstances through the differential survival of *causally unrelated* random mutations (Dawkins, 1997; Dennett, 1995; Cronin, 1991); it became, in the minds of many mainstream biologists, *Adaptation* as a *Final Cause*. That is, adaptation, as an abstract notion, was treated as the cause of evolution. Thus design was brought back in - through the back door. Theoretical developments also began to emphasize 'processes of group selection', sometimes at the level of species, so that natural selection not only seemed to be part of a greater design but also less individualistic and selfish; it was deemed *social* and could be in the direction of *altruism.* Natural selection towards Adaptation (thus seemingly Fulfiling the laws of nature with progress the inevitable result) was emphasized at the expense of sexual selection (sexual desires, sexual competitions, sexual deceits and oedipus problems, for example), so the process seemed rational rather than emotional, sexual or political (see next chapter).

The problem was, thus, solved for most scientists, and sufficient concessions with Darwin's thinking were made so that philosophers, theologians, reformers and educated people generally could be on board. God was back (if wanted), morality was reinstated and progress a possible conclusion of the apparently true fact that humans, the *rational* animal, came at the very end of the evolutionary tree. Unfortunately, back also was the teleology from which Darwin had so desperately sought to escape. By the mid 1960s work began to appear in biology (Cf., Williams, 1966: Hamilton, 1964, 1971; Morris, 1968) which pointed out the extent of the deviation from Darwin which had taken place in the biological sciences. These scholars generally argued for a return to the basic principles of chance in a process of *individual* selection in which sexual selection - and all its blind-to-rationality drives and emotions - was as important, if not more so, than natural selection. This movement crystallized with the publication in 1975 of E. O. Wilson's, *Sociobiology: the New Synthesis,* and Richard Dawkins', *The Selfish Gene* in 1976, in which these processes were to be applied to human evolution just as to the evolution of other species. So, once again. evolutionary biology and the social sciences were on a collision course.

BIOLOGY AND THE SOCIAL SCIENCES: THE SEARCH FOR MORALITY

The evolving social sciences (see note one, p., 234) also worried about the full implications of Darwin's postulations. When evolution as a product of natural selection was (mistakenly) presented as Social Darwinism – a constant struggle among individuals in which only the strongest survived (Cf., Spencer, 1862, 1967; Sumner, 1992) - it was attacked directly (Cf., Ward, 1893, 1897, 1906; Bellamy, 1888; for a discussion see Hofstadter, 1955) as not being relevant to the study of human behaviour. This set in motion the social sciences' long term answer, which was to *separate* the biological from the social and treat them as distinct areas for study and, increasingly, as separate *moral* perspectives

So, biologists and social scientists have, since at least the turn of the Twentieth Century, more or less gone their separate ways. Social scientists (and others labelled as humanist) have increasingly become mistrustful of biology and biologists where explanations of human behaviour are concerned, fearing a sort of anti-humanism, anti-rationalism on the part of biological determinism. Humans, they argued, became, with the evolution of reason, moral agents. So attempts to use biological material in the explanation of social behaviour is not only misguided but potentially immoral. Any practitioner who dared try was considered to be part of a conspiracy, wittingly or unwittingly, to 'maintain the status quo'; to oppress subjected and disadvantaged people, and to be in support of racism or capitalism or some such postulated evil.

This view of biology and biologists by social scientists has not exactly endeared social scientists to biologists, who do not give much credence to the value of the social sciences in the first place. Biologists generally ignore the social sciences, considering them (and often the humanities as well) as not much more than 'common sense', often superficial and boring, and their practitioners more or less modern day myth makers whose work positively stands in the way of scientific progress (Cf., Symons, 1979; Tooby and Cosmides, 1992; Pinker, 1998). There is generally a feeling that social scientists are a bit weak of mind, seeking glory through wild speculations and continual accusations against those designated as perpetuators of social wrongs, claiming the moral high ground, while being incapable of 'real science' or real intellectual integrity. At best biologists, and other physical scientists, allow that humanities and social science are 'the soft sciences'. The result of all this ignoring, suspicion and often hostility, has been to increasingly drive the two disciplines apart, and to feed growing feelings of moral dislike on both sides.

As noted, during the last two decades or so the mistrust of biology by social scientists has been greatly intensified with the arrival of a branch of biology called sociobiology. This is because sociobiology (and to some extent its recent offspring, psychobiology) has dramatically set out to bring back Darwin's original ideas *and*, this is the crucial point, apply them to human social behaviour. In doing so - thus violating the long tradition of leaving explanations of human social behaviour to social scientists - they have directly challenged a number of basic 'moral' and ontological assumptions (and epistemological, as well as theoretical, postulations) found in the humanities and traditional social sciences. The shock and seeming arrogance of this has so stunned a number of social scientists that they have sought reinforcement for their belief that biological explanations of human behaviour are, at best, nothing more than attempts for simple deterministic (thus 'pseudo-scientific') explanations. And at worst, they are portrayed as unwittingly playing into the hands of exploiting, privileged self-interests who want to claim that existing inequalities are natural.

The case of sociobiology is not helped because, even with their interest in human behaviour and claims of having discovered a new paradigm, they have failed to provide a set of concepts, or a theoretical framework, for answering most of the questions social scientists are interested in. It is one thing, for example, to claim superior scientific credentials with the notion that their 'discovery' that the "*ultimate*" cause of human social behaviour is an evolutionary tendency to maximize "reproductive fitness", but quite another to show how that `fact` can be translated, through "proximate" motivators, into, for example, modern social hierarchies or modern political states or a human tendency to moral evaluations. The reason for disappointment is that sociobiologists have singularly failed to develop a conceptual bridge between genes and behaviour (Trigg, 1982) and worse, seem to act as if it were not necessary or important to do so. (For a discussion and an exception see Alexander, 1979, 1979a, b; Trivers, 1981; Lumsden and Wilson, 1981; Dawkins, 1982). There is a sense that doing so would be 'unscientific'; it would be lowering the study to the inexactitudes and 'woolly' concepts of the social sciences.

Although theoretically more interested in the notions of mind and consciousness than sociobiologists, psychobiologists have not moved us much in the direction of discovering their characteristics; in fact it might be argued that they have moved us backward (see next chapter). They have been more concerned to explain the evolution of, for example, sexual competitions, deceitful behaviour, language, altruism and kinship in *mathematical,* computer like (rationalistic) terms rather than as expressions of more recognizable human nature motivators such as lust, love, sympathy, fidelity, jealousy, self-interest, status seeking, and so on. They have also tended to ignore rather more complex social/political/cultural behaviour, such as, for example, vengeance seeking, clientism, social hierarchies, the state, politics and religion (to a significant extent all based on powerful, not always conscious, *emotional* motivators). Indeed, among a number of psychobiologists there has been a return to the 'adaptionist paradigm' whereby it is strongly argued that if adaptation is not exactly the Final Cause it is the major, if not only, cause worth considering in terms of natural and sexual selection (Cf., Barkow, Cosmides and Tooby, 1992; Symons, 1992; see next chapter).

As noted, one is often left with the impression that, for sociobiologists (and to a lesser extent, psychobiologists) dealing in ultimate causes carries with it feelings of moral and intellectual superiority which would only be polluted by an interest in proximate causes. To them, concepts and areas of behaviour such as emotions, ego, self, species being, desires, fears, ideologies, religion, love, envy and politics, for example, seem quite wishy-washy, non-scientific, and designed to relieve individuals of the moral responsibility of intellectual honesty. The use of such concepts is considered evidence of a problem humanists and social scientists have in facing up to the difficult truth about our species - that we, too, are animals. But social behaviour in humans, and in all other species for that matter, is a functional product of proximate motivators, not ultimate causes.

On the other hand, as noted, when biology, sociobiology/psychobiology or otherwise, is mentioned to social scientists they become very uneasy and are quick to dismiss their importance for understanding human behaviour on the grounds that they are "reductionist" and "deterministic" (e.g., Herbert, 1983), if not out and out reactionary. Biology (usually assumed to be some sort of 'biological determinism') does not, they argue, take account of rational man/woman; it does not seem to recognize that humans are planning and goal setting creatures; the human capacity for developing a wide variety of symbolically based cultures appears absent from biological considerations, and above all, biology does not seem to give credence to human free will and the human capacity to seek out, impose and enforce moral behaviour. Yet social scientists have made no more headway than biologists in trying to fill the gap between biology and behaviour. After rejecting Freud's attempt to approach the gap from the social science end, social scientists, including many psychologists, shunned such a concern and turned to a worship of ‘rationality’ (and its corollary, 'learning'), to cultural and/or social determinism - the social constructionist position. Yet we await a definition of rationality and 'socio-cultural constructionism' which is not tautological, circular and/or teleological.

There are, therefore, perceived 'moral' as well as ontological and epistemological difficulties at the precise area of concern where biology and the social sciences have to meet. Often these are considered to be irreconcilable differences on the part of practitioners of the two disciplinary areas involved. This is because both areas are, in large part, claiming that they are onto a 'higher' moral and/or scientific truth than the other. As a result, they most often set out to answer different questions while using different meanings for a number of key concepts. Rarely are there attempts, from either side, to conceptualize human nature in any way which would show a believable link between biology and behaviour. To many social scientists (and not a few biologists) it seems, in fact, that it would be immoral to do so.

If we look more closely at how this split developed and has been perpetuated, we will be able to identify more specifically some of the perceived ‘moral’ obstacles which it can be argued have stood in the way of developing a non-teleological concept of human nature. This will also be to the end of arguing for a moral truce, because if these disciplines continue to be seen as irreconcilable, and the practitioners to see each other as potential perpetrators of immorality, it is not likely that they will listen to, or talk to, each other in terms of the task of using material from both to formulate a scientifically usable conceptual link between biological and social aspects of human existence. However, if we can clear away some of the moral objections it should allow us to look more clearly, and hopefully dispassionately, at some of the key epistemological and ontological differences which have grown up with the development of two separate disciplines.

In what follows sociobiology and psychobiology will largely represent biology. This is because they are the branches of biology which most clearly represent those dimensions of Darwin’s evolutionary paradigm that recently have come into conflict not only with the social sciences but many other quests for understanding the human condition. They represent, at the moment, the theoretical arrowhead of biology in relationship to human behaviour. And given that the interest of this work is to establish a human nature basis for a social science, the emphasis will often be more on sociobiology, because, as an aspiring paradigm, it sets out most directly to challenge the social science monopoly on explanations of human social behaviour; for sociobiologists it is not just cognition, language and group size that has to be explained, it is society itself, along with its connotations of being the guardian of human virtue and morality.

The Search For Morality

Social science, even before its rapid development since the late 1800's, has suffered from a schizophrenic tendency. The major problem has been what Jarvie (1984) has called the problem of "the one and the many" (p. 7). By this he means that there has been a tendency to treat humanity as a moral unity, but at the same time to defend the sacredness and individual integrity of a great diversity of societies and/or cultures - a methodological many. This was part of a developing idealism which had as its major goal the protecting of non-western peoples from exploitation. Historically, when confronted with diverse peoples on the edges of Western expansion, Europeans debated the question of whether or not such peoples were human. Were they of the same species as the white man from Europe, did they have the same biological origins, did they have the same potential for cultural development and progress? These were the questions being debated not only in academic circles but also in churches and government offices (Harris, 1968; Kuper, 1983).

For those who tended to see a diversity of origins (to split the species Homo sapiens into different lumps) it was easy to conclude that there were different kinds or races of humans and that there was, thus, a possibility of natural inequalities, justifying, for example, slavery and caste. The largely unequal economic relationships with 'natives' seemed not only fair but possibly even necessary for the good of the 'inferior races'. In those days, biology, especially evolutionary biology and genetics, was seen as a natural ally of those holding to such views, and was often used as a 'scientific' justification for them, and for the consequent social inequalities that were deemed natural by such advocates.

On the other hand, for those who tended to see a moral and intellectual unity of mankind, the above did not follow. Unfortunately for them, the search for a 'psychic unity of mankind', which was undertaken at about the turn of the century, did not seem to help in discovering such a moral unity. It often implied a Freudian man, a humankind of unconscious passions, oedipal complexes, sexual repressions and death wishes. These were not exactly what those searching for a moral unity of mankind were looking for. Instead it was thought necessary to 'educate' natives into the ways of civilization, thus freeing their 'natural' rationality.

In order to do so, social scientists (specifically anthropologists) considered it important to discover the principles underlying individual cultural patterns and social structures - thus the concern with the methodological many - so that native societies could be helped to progress through the predetermined stages to civilization. To accomplish this it became fashionable to deny the existence of human nature, of unchangeable, uncontrollable human emotions, and to place great emphasis on the importance of environmental influences on behaviour and thought (Freeman, 1984). By the second half of the 20th century, this group was, in academic circles at least, clearly in the ascendance.

The Taming of Human Passions

Work in this vein fitted in nicely with much earlier developments in major world religions and philosophies. In these religions the degree of control of human passions became the basis for moral judgments about good and evil. Philosophers agreed, but also added a worship of reason. (Eventually a number of philosophers and theologians, at least in the West, set out to completely separate mind and body.) Both theologians and philosophers recognized, however, the possibility, indeed probability, of uncontrolled human passions, which they perceived as evil. This was often paramount in their thinking about the nature of good. Evil had to be overcome in order for good to exist, but this was possible with proper rituals, confessions, worship and/or reason/rationality. Social scientists, at least sociologists and anthropologists, did one better; they attempted to do away altogether with the possibility of evil. That is, social scientists did away with the notion of human nature and thus with the existence of passions, controlled or otherwise.

Building on the work of John Locke (1881), in which it was argued that the human mind was *tabula rasa* and that all knowledge was empirically acquired writers such as August Comte (1974) evolved humankind through the mental stages of superstition, metaphysics and then to the final stage of true rationality, to positive philosophy. Positive philosophy was more than a method of understanding the environment: it was a stance that rationality would reign supreme in most, if not all areas of life, just as soon as the pollution of superstition and traditional thought were replaced with objective, empirical (scientific) knowledge. Evil was ignorance, nothing more. And ignorance would soon be defeated by rationality.

Many social scientists (and philosophers) came to argue that human social/cultural evolution was fundamentally different from biological evolution. This was because, whereas biological evolution is about uncontrollable organic changes which take place over millions of years, social/cultural evolution is about the accumulation of knowledge and cultural artefacts during single generations which can be passed on to succeeding generations; that is, it is 'progressive'. Social scientists argued, moreover, that because of the difference between organic and cultural evolution, we have in the latter further evidence of a 'rising above biology', a freedom from the 'iron determinism' - as many people saw/see it - of genes, and an ending of any connection with natural and sexual selection. Nature does not control human destiny, it was argued, but 'rational' humans are in charge of their own destinies, indeed, of humankind's collective destiny.

However, in order to maintain the sense of purity which social scientists attributed to rationality, to the acquisition, as they saw it, of 'Truth' (Morgenthau, 1946), to the denial of human passions and to the notion of progress, in the face of what appeared to be continuing evil and dubious progress, they increasingly began to speak of a theoretical humanity, an idealized humanity of the future (Pocock, 1985; Carroll, 1985), one without evil or passion. For example, as the Twentieth Century unfolded, for economists 'the market' became a 'rationally' comprehensible abstraction, not a place where teeming, screaming, scheming humans bargained fish for pottery, rice for knives, corporate assets for power, and honour for money. At the same time Marx was abandoning his search for "species being" and was concentrating his efforts on examining the contradictions and tendencies to class formation in various types of societies so that societies - and social evolution (revolution) - could be postulated separately from inherent human motivations. In the end Marx attributed rational action to classes (Morgenthau, 1946). Political sciences undertook studies of 'the state' and constitutions as social and theoretical abstractions, and sociologists, as we have previously seen, began to study societies as functional systems.

Thus, in the eyes of many social scientists, suggested linkages between biology and human behaviour represent a re-introduction of evil into the human soul; a desire to argue that the world is not the 'nice', reasonable place it should be, and that, furthermore, it is unchangeable. An introduction of biology into the social sciences would seem to be a regression into the dark days before philosophy, science and, above all, human rationality replaced superstition. Its advocates must, it is more or less believed, be complacent concerning 'evil'. Since 'evil' today is considered by many social scientists to be social inequality, racism, sexism and fascism, sociobiologists and psychobiologists must be in favour of these, or dupes of those who are.

Paradoxically, it is with one of the most abstract schools of the social sciences, structuralism, that sociobiologists are most in sympathy (Cf., Wilson, 1975; Alexander, 1979a). Upon examination this is not so surprising. It is not, Alexander tells us, that all behaviour must be predictable from fitness theory, but rather that social "forms" as abstracts fulfil what fitness theory would expect. This use of structuralism allows sociobiologists their ultimate cause (inclusive fitness maximization), their population genetics and their theoretical postulations about principles of sexual selection, without their needing to consider proximate causes of behaviour. Sociobiologists too are not averse to escaping the problem of analyzing human passions. They too worship rationality.

It may be understandable that sociobiologists like this brand of structuralism's return to a psychic unity of mankind. They must realize, however, that social science structuralists are talking about unique 'rationality' and language as structures of 'mind' acting as determinants of behaviour; sociobiologists, in the end, are talking about genes as determinants of behaviour. This difference has been perceived by a number of social scientists as, in fact, bringing out most dramatically the continuing nature of the negative use of biology in earlier attempts to explain human behaviour; a use which social scientists perceived as meaning that behaviour was unchangeable, that, perhaps, the races were fundamentally different, forever and ever, that white men (and women) from Europe were superior to all other humans. The apparent conflict between mind and language, on the one hand, and genes on the other, suggests to social scientists straight genetic determinism and a re-postulated essentialism, which could again be used as justification for what now seems to many to have been unacceptable hierarchy and exploitation. Only now the victims are thought to be peoples of the third world, women, members of minority groups and the poor.

In reality, however, social scientists were never really able to get rid of human nature. They only changed it from being biologically based to being 'morally' and 'rationally' based, with learning, sociability and a ‘natural’ tendency to cooperation postulated to play a major role in determining individual behaviour. And it is this picture of humanness that the notion of the selfish gene, ‘designed’ only to replicate itself, reeking of uncontrolled selfishness, challenges most dramatically. A large part of the conflict between biologists and social scientists, then, is a conflict about the fundamental ‘starting morality’, or lack of it, of basic human nature.

Basic Human Nature

Are humans basically competitive or cooperative, cunning or open, selfish or altruistic? That is, are humans basically bestial or spiritual? Sociobiologists provide a framework which suggests that humans must, at base, be competitive, cunning, selfish and bestial. On the other hand, social scientists have spent long years trying to prove that humans are basically cooperative, open, altruistic and spiritual.

The conflict here would seem to be very clear. Yet it is not. Sociobiologists recognize the existence of cooperation, openness, altruism and human spirituality. Their response has been both illuminating and disappointing. It has been illuminating because they have gone a long way to show, theoretically at least, how a number of apparently cooperative, open, altruistic and spiritual acts can work in the end towards the perpetuation of certain individually carried genes (Cf., Hamilton, 1963, 1971; Williams, 1966; Trivers, 1971; Alexander, 1979a, b; Crook, 1980; Barash, 1982; Ridley, 1993, 1997; papers in Cronk, Chagnon and Irons, 2000), as, for example, in parenting, altruism, kinship behaviour and behaving as a nice person. It has been disappointing, though, because sociobiologists (and psychobiologists) very often seem to refuse to grant *current significance* to motivators which do not appear to directly or theoretically contribute to reproductive success, or to adaptation during an earlier stage of evolution (Cf., Symons, 1992; Barkow, Cosmides and Tooby, 1992).

Take, for example, a strong desire for status. This could have given males reproductive success if status added to "attractivity" during a very early period of differentiation (Symons, 1979) but could detract from reproductive success in an age of contraception and possibilities for career advancement and/or high material life styles. That is because in such circumstances individuals and families often decide *not* to be burdened by too many offspring (Lopreato, 1984; eds., *Biology and Society*, Vol. 5:1, March, 1988). However, the observation of reduced reproduction rates among career and status oriented people, and the fact that status seeking is generally considered in modern times, especially by scientists, to be *non-rational* - if not self-centred and selfish - means that it is generally not considered a major aspect of human nature for sociobiologists or psychobiologists (and most social scientists). The fact of potentially reduced reproductive success in modern societies, however, does not make the *desire for status* any less an aspect of human nature, and therefore a motivator of behaviour, than when it was more directly beneficial to reproductive success (and could have been argued to have been an adaptation). The social outcomes might be different, but that is what a science of behaviour is about, explaining human social outcomes.

To take another example, a high male sex drive can lead to a long life of reproductive success in some circumstances but in another context it can lead to Aids. It nevertheless remains a powerful motivator. The *social* outcome of this desire, however, clearly varies with circumstances; moreover, outcomes may become quite complicated. For instance, a strong male sex drive may lead to a lot of girl friends and a playboy lifestyle in one context, but in an Aids epidemic can result in – in conjunction with certain other emotions, desires and fears – ambiguity, mental conflict and anxiety concerning sexual desire. These can result in what we call *guilt* (see Chapter Five) induced by the onset of sexual desire, which in turn can lead to advocating and/or accepting all sorts of puritanical and *moral* injunctions concerning sexual promiscuity. But *motivations* (and struggles) *to control sex drives* (which may, in fact, have been favoured by selection at an earlier stage of human evolution because it motivated behaviour which did not scare off potential mating partners, alienate potential male alliance mates or alert rival males) play little part in explanations of human sociability for sociobiologists and psychobiologists. It is all a bit messy; none of this is easily linkable to genes in the way that 'sex drive' treated as a mechanical force is; it does not seem to be rational, calculating behaviour. It smacks of sublimation, projection. For most biologists guilt is decidedly beyond the pale; it affects, they feel, the weak-minded. Guilt is certainly not something from which a *real* science can be built.

And it does not lend itself to the notion of *genetically created moral* categories of behaviour (as, for example, 'genetically induced altruism' or kinship do). Sociobiologists and psychobiologists too are interested in moral 'man' but do not want to go through any non-rational, messy, slippery, non-cost-benefit analysis, mathematically unfriendly concepts to get there. Sociobiologists and psychobiologists, in fact, often stick with their moral categories whatever the evidence. For example, kinship behaviour might be an advantage in some circumstances (where, for example, there is a strong desire to defend fixed property over a period of time), but a disadvantage in others where the free movement of *individuals* and opportunities for friendship and ad hoc alliances might provide greater reproductive success. And given the lack of significantly extended kinship in hunting and gathering societies (see note two p., 234), and in modern societies, it is, therefore, questionable, whether we have powerful motivators for such social formations. But the hypothetical concept of kinship is very convenient for explaining, in cost-benefit terms, sociability in the context of the selfish gene and individual selection. So, sociobiologists and psychobiologists have generally embraced it with enthusiasm.

Kinship, it can be argued, is used by sociobiologists and psychobiologists in explanations of human social behaviour because it seems to be a 'good' - rational and moral - outcome of the selfish gene, and suggests a moral society, rather than because it is derived from observations of human behaviour. What it amounts to is that sociobiologists and psychobiologists have often sought to keep their selfish gene intact but allow for sociability by creating 'moral' categories (for example, parenting, altruism, kinship, reciprocity). As a result, they have tended to ignore many of the less rationalistic concepts developed by social scientists which might, in fact, be useful in linking biology and behaviour in societies (Fox, 1986), and in explaining general patterns of behaviour. Biologists (psycho and socio) need to accept that their perspective can not demonstrate that a given set of genetic propensities selected in one context *must* have the same manifestation and outcome in all or become unimportant for scientific explanations. During evolution mutations accumulate and environments change, but that does not mean that earlier motivations do not persist, sometimes as a hindrance to reproductive success, but not sufficiently so to end the flow of particular genes; certain motivators can become neutral with regard to reproductive success but manifest themselves as other behaviours, and so on. Success, genetic or social, of course, is not guaranteed. Individual organisms die

(Alexander, 1979a), social patterns disappear and whole species often become extinct (Lewontin, 1978; Mayr, 1978).

To be fair, socio and psychobiologists have often stated that an early evolution of 'successful' reproductive motivators could later become counter -productive in terms of a given individual's reproductive success. But what they have been less keen to do is to search for and theorize about such motivators as they could be manifest in a modern context. This might not be their job. But then they should not condemn those who try to find a consciousness which has some basis in evolutionary theory and can be linked to other dimensions of human nature, which in turn can be seen as motivators of behaviour in modern societies. Sociobiologists and psychobiologists have tended to treat the selfish gene as a rational calculator, then have linked it to a number of moral categories of behaviour (while rejecting immoral or fuzzy categories, e.g., oedipal struggles, guilt based reactions, fratricide, vengeance seeking); the result has been rational and moral social behaviour. The search for rationality and morality is hard to escape, even by those explicitly coming from a supposedly amoral, scientifically hard headed Darwinian perspective.

Likewise, the work of social scientists has also been both illuminating and disappointing. It has been illuminating because such things as self-esteem, status-striving, status-defending, kinship behaviours, vengeance seeking, social conflicts, power relationships and politics, for example, have been described and analyzed in great detail. It has been disappointing because these have not become the basis for discovering a consistency of human nature for use in explaining other social behaviour, or even for looking for patterns of their reoccurrence. Rather they have been used to explain, teleologically, the Fulfiling of either social systems or utopian goals; or as lessons for improving behaviour; or as behaviours not to be encouraged; or as examples of human bouts of irrationality. Whatever the case, they have been analyzed to argue that humans are rational animals who can do anything they set their mind to. Social systems and culture are to be created, each one can be that little bit better than the last. This has driven their work further and further away from what they consider to be the morality antagonistic zone of 'genetic determinism' or a concept of human nature.

So, all of the recent discoveries and theoretical developments in ethology, genetics, biochemistry, neural sciences, psychology, cognitive sciences and evolutionary theory have not brought fusion. Scientific observations have not overcome moral hostility and distrust. I suggest, again, that it is because the emotional impact of the 'moral' ambiguity and confusion in-troduced by Darwin is so very profound, and it has not gone away (it may never go away). Currently, and probably most significantly, such ambiguity and confusion are becoming especially manifest because of threats to the very separating developments mentioned above. As long as the two disciplines were (are) separated, and potential linkages un-investigated, the issues of purpose, direction, design perfectibility and progress could (can) exist as mere assumptions (by both sides). These notions do not have to undergo philosophical or scientific scrutiny.

However, theoretical developments in biology are increasingly looking to fusion within the discipline itself, progressively diminishing the idea that clear distinctions can be drawn between areas of biological activities, including between brain and mind. Bio-chemistry, physiology, anatomy, ecology and genetics no longer 'naturally' develop on their own; their very success has increasingly led to an integration, through, in significant part, the use of Darwinian evolutionary thinking. Most significantly, these disciplines are increasingly being used in explanations of mind and behaviour. Moreover, such developments as the human genome project and genetically modified food (and organs) have brought the potential impact of genes and genetic selection directly into the consciousness of humans from all walks of life. And so these developments are leading to an ever increasing sense of invasion of biology into human social and moral life.

The nightly news, documentary programmes, stories in popular magazines and pub conversations are full of the 'miraculous' powers for both curing and potential Frankensteins of genetic engineering, of the mystical powers of the human brain (not mind), of scientists in white coats designing our babies. Similar dramatic, tantalizing documentary and nature magazine presentations of foreign and exotic cultures, far away, which in the past so intrigued the Victorians and later the listeners of the Home Service on the BBC, bore us. Our own culture, at the margins, and in the media, is full of even more and 'better' weirdness (weirdness that probably no previous culture in human history would have tolerated or thought of).

Significantly, neural and cognitive sciences have not only invaded psychology, as noted, but these invaders have increasingly looked to evolutionary theory as their unifying framework. The impact of evolutionary biology as the potential scientific 'lunge forward' manifests itself once again. It is almost as if it was Darwin's day anew; the fixity of species (and with it some of God's power and monopoly on design) was forced to give way before Darwin. Now, not only is God challenged, but so too are all the 'scientific developments' to create rational, thinking, socially constructing 'man', and to bring back purpose, direction and moral progress, which for a time Darwin seemed to threaten. And with this challenge, the full brunt of old Darwin returns (of natural selection and sexual selection being based on random mutations, the randomness being causally un-connected to outcomes in any design, purpose or progressive sense - of Fortune and Nature being all powerful). These seemingly stark possibilities are harder and harder to ignore by ordinary people, let alone by biologists or social scientists. The ambiguity and ambivalence of both the threat and promise are no less than during Darwin's expositions. For many the question still exists, 'is God in favour of all this new science, and its potential impact?'; or 'are scientists running wild, out of God's control?'; or 'is there no God, only mad scientists?'. Is human morality under grave threat? Darwinism, it seems, once again 'needs' to be brought under control.

The result of this renewed threat and ambiguity is that, while it is increasingly difficult to maintain a separation between biology and behaviour, this very fact (which runs counter to the long history of development in both biology and the social sciences) generates an even stronger *compulsion* to make every effort *to do so*. As we have seen, much of the impetus behind the origins of the social sciences was, in fact, to separate the study of human behaviour from biological interference. 'Biology as an enemy' became part of the ethos, if not definition, of the social sciences. The more observations, discoveries and theories threaten this, the more social scientists dig in. For their part, biologists have busied themselves with anatomy, physiology and looking for the *adaptation* of species other than human; there was plenty of work here without bothering with humans; biologists could be 'scientific' because they were seeking to understand the laws of nature not un-measurable things such as consciousness or beliefs or self, for example.

But there is a danger for biologists in this; Darwinian oriented social scientists (there are a number) might grab the ground between biology and mind or, more threateningly, between biology and behaviour. Social scientists may take over all the interesting biology: the biology of consciousness, of thinking, of sociability, polluting it with their 'do-good-ing' concepts. So biologists hang on to the divide by fiercely defending what seem to be purely biological concepts – such as a slightly disguised genetic determinism (for example, arguing for gene-culture co-evolution via 'culturgens', or postulating genes for *abstract categories* of behaviour, such as altruism), or putting forth adaptation as *ultimate* cause not really requiring any understanding of proximate mechanisms (see next chapter). When they have looked outside of biology they have not looked to traditional psychology or to the social sciences, but in what they conceive to be the opposite direction - to mathematics and computer science. They do not consider themselves genetic determinists, but rather they want to be known as 'real' scientists, at the forefront of understanding mind as a rationality machine, generating a calculating science which brings nature under human control; humans need not be left to the wiles of guilt, on the one hand, or the prejudices of ill conceived do-good-ism, on the other.

So, neither the new opportunities born of current scientific developments, or their threats, seem to bring biologists and social scientists together; rather these seem to drive them even further apart. In reality, however, it is very doubtful that there are overriding epistemological or ontological reasons why concepts from biology cannot be linked to those from the social sciences (see next chapter) in search of a basic human nature. There are not *two* humannesses, one to be studied by biology and the other by social science. There is no reason to support the view that the bogus biological justifications used in Nineteenth and early Twentieth Century racism and colonialism, to designate some humans as being innately inferior, should condemn human social biology as a science; or that the 'evils' of the supposed support for modern racism, sexism and market economics (by designating humans as being basically selfish), ascribed to biology in more recent attacks on sociobiology, can be attributed to some intrinsic nature of biology, *qua biology* (Ruse, 1979; Hull, 1978, 1978a; Trigg, 1982; Runciman, 1998; Pinker, 1998; Gray, 1999; Brown, 2000).

At the same time, there are no grounds to believe that the social sciences are in some way more moral than the biological sciences because social scientists are often studying issues related to human welfare (in so far as medical sciences are based on biology, so too are biologists studying issues concerned with human welfare; and anyway, how do we know what the attitude of any given biologist is to human welfare – biology, as a discipline has no view on welfare). On the other hand, it is not the case that the biological sciences are morally superior because they advocate a hard hearted, *realistic* approach to science and an understanding of human behaviour unclouded by ideological preoccupations. If we are to come to grips with 'basic human nature' as a means to progress a science of human behaviour, biologists and social scientists must work together, or at least the disciplines will have to be brought together. Both disciplines have accumulated a great deal of knowledge which will be essential to our search for a non-teleological human nature; more importantly, human nature is by definition a fusion of the two areas. It would be a shame if those who already have expertise and crucial knowledge did not participate.

A contribution of sociobiologists and psychobiologists could, for example, be to help uncover proximate mechanisms - human nature motivators - by extrapolating them from principles of sexual selection as applied to an earlier period of human evolution. They could also use these criteria to check the human nature processes being suggested from other quarters – such as from the neural and cognitive sciences, for example. The job of the social scientist could be to use these human nature processes to help explain observable behaviours and social patterns in a variety of historical contexts and to suggest human nature propensities based on discovering universal patterns of behaviour. To return to an earlier example, a universal desire for status, or more specifically a desire for positive recognition of and attention to self, could explain a considerable amount of what appears to be human altruism, reciprocity and general sociability. Whereas status-striving as a non-rational behaviour or as being dysfunctional for social systems; or 'reproductive success' as a cause of altruism, and altruism as a cause of altruistic behaviour, seem either meaningless or clearly contradicted by observations of much of modern behaviour.

The distinctions set at the beginning of this section are probably false distinctions. Humans are neither basically competitive nor cooperative, cunning nor open, selfish nor altruistic, bestial nor spiritual (see also, Boehm, 1997). These distinctions are as much a product of hostility between two disciplinary areas in which practitioners have looked to distance themselves from each other as from any intrinsic nature of the subjects being studied, or from any scientific 'truths' privy to one or the other of the disciplines. Without enlightenment assumptions about rationality, perfectibility and progress, observations (and evolutionary theory) might suggest that instead of one or the other of the binary opposites above, humans are more likely to be self-aware, anxious, evaluating creatures who desire to: protect self (dignity), maintain prestige (social acceptability), defend rights (status) and control their physical and social environments. And in their social interactions they are very likely to come to think that they and the universe have a purpose. This, however, brings us to another major area of moral resistance to a concept of a biologically based human nature, even among biologists.

The Problem of Human Purpose

As noted in Chapter One, most humans like to think that they have a purpose, that their life has greater meaning than simply the satisfactions of a moment - although its fulfilment may be somewhere in the future. Among those who study society and history there are many who like to think that there is a purpose and a direction in human existence. This seems to be no less true of sociobiologists (and psychobiologists) than it is of social scientists. Otherwise why would sociobiologists and psychobiologists almost invariably end their writings with some warning to the effect that the only way to advance human prospects, and to avoid certain tragedies, is to know our true (biological) self?

Yet if we take the sociobiological framework to its limits, there is no individual, social or historical purpose, separate, perhaps, from maximizing reproductive fitness; but even this is a non-teleological, random process. There certainly is no purpose in the direction of social evolution as derived from the full implication of this paradigm (but see, Barkow, Cosmides and Tooby, 1992; Symons, 1992; Dawkins, 1997). Indeed, taking the sociobiological perspective to its extreme, notions of purpose are only products of genes, and, although genes are variable to a considerable degree in their outcomes, individuals or groups (even sociobiologists) cannot overcome the genetic programme of purpose set for them. Social scientists trying to comprehend the sociobiological position, thus feel, with some justification, that biologists are rejecting the existence of civilization, and with it the possibility of a higher purpose. This further makes it appear to social scientists that sociobiologists (and psychobiologists) are not only reactionary in terms of the notion of progress, but also that they have not noticed that humans can, and do, construct higher purposes and work hard to put them into practice.

Biologists, it appears, are in a dilemma of sorts. They do not seriously feel that they are reactionary. Indeed, they consider themselves to be at the forefront of a scientific revolution. Nor do they believe that humans cannot control at least some aspects of social evolution. But to embrace this would greatly weaken the claims of their paradigm; thus their protestations about the importance of human knowledge, understanding and improvement being dependent on biological understanding. But biologists have another approach as well - to hide in the sanctuary of science. Because they work and theorize about bio-chemical units (genes), can use mathematics effectively (population genetics) and understand the non-teleological principles of natural and sexual selection, allowing them to make many meaningful statements about animal behaviour, they assume, rightly, that they must be onto something important when it comes to human behaviour. Therefore, they often feel that criticism from social scientists is not worth considering.

Furthermore, they feel that what appears to be the often mystical human ethnocentrism of social scientists is due to a lack of scientific rigour inherent in their discipline, and sometimes in the personalities of those who practise it. This, they feel, leads to an unwillingness to face the 'truth' about the universe of humans as being subject to the laws of science. This apparent arrogance, however, further convinces a number of philosophers and social scientists that science itself is really only a cultural artefact, its truths relative only to prevailing paradigms and 'dominant' cultural values, quite separate from universal knowledge, let alone from consideration of significant human problems.

Reinforcing this view, social scientists look with pride at the work during the latter half of the 19th century and the first half of the 20th century with its apparent richness in achieving knowledge which seemed to contribute to human advancement and progress (Nisbet, 1969; Kumar, 1978). However, if we take seriously the arguments of most social scientists, especially those of a reformist nature, we get only a very vague picture of human purpose. It somehow has to do with assumed virtues of social stability and/or the search for 'progress' through the development of new technologies, improved divisions of labour, 'proper' systems of social justice and a general spread of democracy (presented in the abstract rather than in detail). All of these might be worthy concerns but they hardly represent a 'value-free' method of discovering human purpose. Furthermore, social scientists argue a great deal among themselves as to the 'true' nature of progress, justice and democracy, leaving the impression that their notions of purpose represent nothing more than their own prejudices and preconceived hopes.

Morality, Biology And A Social Science

Given all this, despite the fact that there are clearly a number of emotionally based 'moral' objections in the way, I see no reason why it is impossible or undesirable to proceed with an attempt at the integration of biology and social sciences to develop an empirically based science of social behaviour. As noted above, an essential starting point will be for them to work together to uncover 'basic human nature'. I have not been able to uncover any convincing 'moral' reasons why it should not be done or is not possible. Those objections discussed above have more to do with universal human anxieties about human purpose in the universe and/or a desire to separate social, cultural and moral explanations from physical explanations (by both biologists and social scientists) than anything intrinsic to humans or evolutionary processes which would fundamentally stand in the way of integration.

The criticisms of biology by many social scientists, for example, are largely unfounded and often misguided. In the first place, while biology has been used on several occasions in the past to justify what now seems 'evil', it has also been used to fight for the same 'positive' ends that liberal philosophers and social scientists have been interested in; for example: 1) the struggle by biologists and physical anthropologists in the post Second World War years to 'prove' that, biologically, there was no such thing as 'race' and that all humans were the same in terms of all 'significant' characteristics - there was but one 'family of man'; 2) the work of post Second World War primatologists and human evolutionists to explain human evolution, interestingly described by Donna Haraway (1989) as almost a "(Judeo-) Christian Science". This was because it was about coming out of the Garden of Eden, about gaining knowledge, about distancing ourselves from animals, about separating mind from body - above all it was about putting us at the top of the evolutionary ladder, and in the process separating culture from nature, civilization from savagery, gender from sex and race from biology.

What happened was that during the 1950s UNESCO presented the world with a number of papers which, along with 'The New Physical Anthropology', as it came to be called, focused study and theoretical consideration on behaviour as a functional means of biological adaptation in the evolutionary process - 'social functionalism'. As time went along, chemists, geologists, psychologists and social anthropologists were brought in to what was quite an extraordinary degree of inter-disciplinary work. This work was very well funded by the Wenner-Gren Foundation for Anthropological Research, the Carnegie Institute, the American Museum of Natural History and UNESCO, as well as by a number of universities. The result was a picture of humans in which 'the First Human', 'Man the Hunter', emerged, from the 99% of human history spent in gathering and hunting, as a male centred, probably nuclear family oriented, cooperative (with insiders), competitive (with outsiders), potentially extremely rational, liberal (in terms of human rights), indeed anti-racist, creature (Haraway, 1989). Biology, in and of itself, clearly does not always result in 'bad man' (and it is certainly no guarantee against teleology and mythology creation!).

In the second place, while biology in the past has been used to justify what later came to be seen as 'evil', it appears to have been much less used than: ethnicity, religion, status differences, supposed immoralities and social and political ideologies. In fact, it might be argued that biology (usually concepts of different sexualities, 'freakishness', 'race' or differential IQs) is only used when the above are in full flood. The 'evil' perpetrated on the basis of these *cultural* notions, moreover, has been, and continues to be, almost in-comprehensible in terms of its magnitude and potential for intensity. Indeed, when we consider religion, ethnicity, a variety of cultural beliefs and political ideologies, biology as a justification for killing and incarcerating people almost fades into insignificance; if we did not have the example of Nazi Germany, it would. Nevertheless, while

> "The Nazis set the historical standard for efficient mass killing, yet even their record has been broken. Their body count has been surpassed by the Chinese Cultural Revolution of the 1960s. The Cambodians of the 1970s destroyed a larger percentage of the population. The Rwandese genocide of the 1990s killed people at five times the rate of the Nazi death camps. . ." (Baumeister, 1997, p., 383).

Killing in these examples was largely justified on the basis of ethnicity, status, and above all, political ideological criteria. It is worth noting that two of the societies in which the highest number of individuals have been killed in modern times - the old Soviet Union and China (with an estimated 40 million killed between them, Baumeister, 1997, p., 112; see also, Conquest, 1971) - had (and have) very distinct official ideologies which proclaim that all humans are biologically the same. All official killing was done under the guise of eliminating political and ideological imperfection. Yet, those who most vigorously oppose biological criteria in explanations of human behaviour usually are the strongest advocates of some form of cultural or social determinism as the moral way forward; the very factors used in the above behaviours. And, as was noted in Chapter One, advocating rationality does not necessarily mean that the above behaviour will not take place. Indeed, it is often used as both justification for eliminating the irrational – read, non-believers - and in establishing efficient means for their incarceration, mind-altering treatment, re-education and even elimination.

SUMMARY AND CONCLUSIONS

In this chapter I have suggested that although Darwin shocked the world with his non-teleological approach in which design, direction, progress and purpose were taken out of nature, it should not, therefore, mean that biology be banished from discourses concerning human behaviour. It does not mean that the enlightenment enthusiasm for finding design, direction, progress and perfection in humanity justifies taking humans out of Mr. Darwin's nature. Unfortunately, the history of moral differences described in this chapter have led to (or at least have reinforced) a number of epistemological and ontological differences which, to some, make it seem that, even if they wanted to work together, it would be impossible, or at least, not scientifically fruitful. I hope that this is not the case, as an examination in the next chapter of the epistemological and ontological issues between them will show.

CHAPTER THREE
BIOLOGY AND THE SOCIAL SCIENCES: EPISTEMOLOGICAL AND ONTOLOGICAL ISSUES

Both epistemological (*on what basis can we claim to know)* and ontological (*what is the fundamental nature of being)* conflicts between biology and the social sciences are manifest in at least three major areas of concern; they are: 1) the problem of causation, 2) the problem of human nature (with specific reference to concepts of 'pre-knowledge') and 3) the problem of classification (deciding upon scientific significance and unity). Each will be considered in turn.

THE PROBLEM OF CAUSATION

Biologists, following Mayr (1961), tell us that in terms of evolution by natural and sexual selection there are two distinct types of causes: ultimate causes and proximate causes. Ultimate causes are held to be final, universal 'processes' which are not affected in any fundamental way by proximate causes. Proximate causes are those 'immediate' events which directly affect other events. Social scientists have not tended to explicitly separate causes in this manner, but it will be useful to analyze their approach to causation using the same distinction. First let us consider ultimate causes as approached by the two disciplinary areas.

Ultimate Causes

Among biologists, sociobiologists make it clear that they are primarily interested in discovering ultimate causes (Barash, 1982). These, we are told, cannot be understood as a conglomeration of proximate causes (Alexander, 1979, p., 225). But what do these ultimate causes affect? Do they cause individual human behaviour, specific human social organizations and a designated direction of human history? Well no, not directly anyway; but ultimately, well yes, they must. Or at least they set the parameters within which all of the above takes place. But what does this mean? When we look at the major ultimate cause presented by sociobiologists it is nothing more than a tautological statement. It is to the effect that genes which are most represented in a given population are there because they provide (or at least have provided) the greatest degree of relative 'inclusive fitness'. Inclusive fitness represents a variety of behaviours which result in the survival and reproductive success of individual organisms (or at least genes); and that is all there is to it, metaphysically speaking.

For those with a conventional sense of causation - in which, when we have A and it has been observed that we seem inevitably, thereafter, to get B, we treat A as a cause of B - this is not very helpful. Nevertheless, the concept of inclusive fitness is very important in that it tells us what probably is not. It tells us that organisms, species, animal societies and evolutionary histories have no pre-determined paths; that the attribution to them of teleologically derived needs (often treated as causes) and purpose (progress) has no place in the study of Darwinian principles of selection, population genetics, ethology and ecology.

In its detail this framework has drawn our attention to the importance of looking for unity and complexity in inheritance while allowing that minor modifications, over time, result in diversity (Cronin, 1991; Dawkins, 1997). It allows for what appear to be rather intricate adaptations to exist in the light of random, accidental modifications, without needing to call upon ultimate design. It also allows for non-adaptive behaviour to exist. In other words, it rids the study of animal evolution and behaviour of spirituality and teleology; it opens it up to the same kind of scientific investigation that bones and organs have previously been subjected to. The tautology of inclusive fitness represents a unity of observations of nature, worked into an elegant theoretical model (neo-Darwinism). This framework has withstood vicious attacks and the test of time to provide innumerable hypotheses and research programmes (Goldsmith, 1991; Brown, 2000), generally without being teleological.

However, it tells us very little about what *causes* the evolution of particular genes, let alone about how these genes translate their messages into species typical behaviour. It tells us absolutely nothing about motivations/propensities for behaviours, such as, for example, the all-important reproductive behaviours of individual organisms. It certainly tells us very little about the motivators of human *social* behaviour. The emphasis on reproductive success challenges the enlightenment inspired notion of what evolution, including human evolution, is about, but is not a discovery of causes. It is an `in the last analysis` statement about the absence of predetermined direction and purpose in the universe of organisms. Social scientists certainly need to pause and ask if the same approach might not be needed in the study of the human species, including its social behaviours and histories. But they will not find causes, as such, in the concept of inclusive fitness.

Classical social scientists too have an ultimate cause. It is 'rationality'. Rationality assumes a universe which is understandable and predictable in terms of functional laws (causes and effects, means and ends) and postulates a human ability (reason), unique among living things, to discover and understand and use these laws. Often it is believed that this 'greater rationality' means that inevitably human civilization is moving in a pre-determined, *progressive* direction - despite certain mishaps along the way. It may take time for reason to overcome superstition, tradition and certain human passions, for example, but that it will do so eventually is rarely questioned. Societies may come and go, but history itself is inevitably purposeful; its purpose is to fulfil its own Greater Destiny. So, rationality as an ultimate cause has as essential ingredients notions of progress and purpose; and a history that is, ultimately, directional.

However, in practice in the social sciences, rationality as a means to understand such phenomena as, for example, human 'needs', 'progress', reasons for (causes of) specific behaviour and/or particular social formations, usually only manifests itself in the manner of circular teleologies (Rosenberg, 1981; Trigg, 1982; see Chapter One). Rationality as something more than an analytical capacity (reasoning and 'instrumental rationality') has not been (and probably cannot be) postulated separately from stated aims and these cannot be separated from desires and/or ideological and culturally prescribed individual and social 'needs' and goals. To take a rather drastic example, it is only 'rational' for an individual to eat in order to stay alive so long as that individual *desires* to stay alive.

Rationality is even less useful as an ultimate cause than the inclusive fitness of the sociobiologists. This is because it does not even clear the way for a science based on non-teleological cause, but rather retards the possibility by enshrining teleology. So, if ultimate causes are of little value, what about proximate causes: might they provide a basis for reconciling the two disciplinary areas? Unfortunately, the conflicts between them are no fewer here than they are with the question of ultimate causes.

Proximate Causes

In practice, biologists and social scientists are as often in conflict over the professed *value* of ultimate as opposed to proximate causation as over the actual nature or 'working' of either. Sociobiologists and psychobiologists frequently try to avoid proximate causes by retreating to their 'fitness' and/or 'adaptation' ultimate cause. This allows them to claim to be practitioners of a higher, more pure science than those (such as social scientists and humanists) more interested in (potentially more messy) proximate causes (see Chapter Two). Social scientists, on the other hand, often concentrate on proximate causes (with little reference to any notion of ultimate cause) in order to highlight the importance, as they see it, of human reason and *learning,* indeed, of humans as moral actors, in social behaviour. However, as noted earlier, the arrival of sociobiology has challenged this stance. At the same time, sociobiologists and psychobiologists have not been able to completely ignore proximate causes, especially when social scientists attack them for being, for example, 'genetic determinists'.

For a few sociobiologists genes themselves are, in fact, the proximate mechanisms of all behaviour (e.g., Wilson, 1975; Dawkins, 1976, although recently Dawkins has backed off somewhat from that position - Dawkins, 1982, 1986). Nevertheless, at a fundamental level this is the basis of the sociobiological argument. It is one that sociobiologists must hold to if their paradigm is not to lose its force (Rosenberg, 1981; for a discussion of how easy it is to interpret their work as genetic determinism see, Brown, 2000). This, however, is a straight reductionist genetic determinism which few sociobiologists overtly wish to proclaim (Cf., Alexander, 1979; Wilson, 1975; Trivers, 1981). This is partially because the relationship between the ultimate cause of inclusive fitness and the operation of its 'selfish genes' (Dawkins, 1976) as manifest in human behaviour, is far from clear and, in fact, seems to violate evidence from both common sense and the social sciences.

At best, selfish genes seem to act in mysterious ways; to list but a few examples: humans practise celibacy, homosexuality, contraception and planned reductions in family size - not only in the most difficult economic and physical circumstances, which is what fitness theory would predict, but, possibly even more often, in the most advantageous of economic and physical conditions. One needs to know little sociology to know that in modern societies the more affluent middle classes have fewer offspring than the less affluent lower classes. It seems that with prosperity the selfish gene becomes less selfish (see also, Lopreato, 1984; Runciman, 1998).

In addition to such problems, it is extremely difficult for intelligent human beings to believe that their observations of a multitude of behaviours and cultural traits, which change over time even within the mind and behaviour of a single individual, have direct hot lines from genes. Moreover, genetic determination, without intervening mechanisms, violates even biological understanding about the interacting ways in which genetic instructions are translated into morphologies and behaviour (Cf., Washburn, 1963a; Lewontin, 1974, 1978; Gould, 1977; Ayala, 1978; Harris, 1980; Ridley, 1999).

As noted, many socio and psychobiologists would, in fact, deny that they are genetic determinists. Instead they often use implied 'needs' of reproductive fitness maximization to teleologically jump from genes to descriptive concepts (for example, kinship, altruism), as proximate causes. They would point to the extensive literature produced by sociobiologists on "kin-selection" (Cf., Hamilton, 1963; Barash, 1982; Dunbar, 1996) "reciprocal altruism" (Cf., Trivers, 1971), "reciprocity" (Cf., Alexander, 1975, 1979a, b) and "altruism" (Cf., Alexander, 1979a, b; Crook, 1980; Badcock, 1986) to argue that these are clear examples of proximate mechanisms which underpin human cultures. For social scientists and philosophers, however, these are largely descriptive concepts (or categories) and as such cannot be used as general proximate mechanisms of behaviour or as an explanation for cultures.

This use of descriptive concepts as causes is epistemologically untenable because it avoids the problem of showing how genes translate into the behaviours described, despite pretending to. To be fair, however, sociobiologists know this and they know that these concepts only represent categories of behaviour for each of which there are a number of proximate motivators at work. But this is not the way it appears in most of their writings, where it would seem as if sociobiologists were arguing that altruism, for example, exists because of the evolution of genes for altruism. This simplification does nothing to further our understanding of the biological and/or psychological bases of altruism, nor of the complexities of relating the selfish gene to altruism.

The problem exists, one could argue, because of sociobiologists refusing to face the crucial problems involved in discovering working, proximate mechanisms. Moreover, in order to marry these descriptive concepts to the individually centred, `selfish` pursuits of reproductive fitness maximization, sociobiologists have had to change the meanings of these words as used by dictionaries, philosophers and social scientists (Barnett, 1983) to mean, for example, not really altruism but selfish-altruism, at best confusing and at worst misleading (see also, Lopreato, 1984).

In more recent times there has been a degree of recognition of this problem among, at least some, psychobiologists (Cf., Papers in Barkow, Cosmides and Tooby, 1992; Dennett, 1995; Dunbar, 1996; Pinker, 1998). A number of these have set out to search for the "missing middle", for an understanding of the ". . . architecture of our evolved psychology. . ." (Cosmides, Tooby and Barkow, 1992, pp. 6, 3; see also, Symons, 1992). But, this work has very often downplayed the earlier sociobiological concern with reproductive competitions and has, instead, looked to 'mind-as-computer'; mind as a problem solving mechanism. "The rise of computers and, in their wake, modern cognitive science, completed the conceptual unification of the mental and physical worlds by showing how physical systems can embody information and *meaning*" (Tooby and Cosmides, 1992, p., 20, my emphasis). For Steven Pinker (1998), equal to evolutionary biology in scientific significance during the 1950s and 60s, was "the cognitive revolution . . . which explains the mechanics of thought and emotion in terms of information and computation . . ." (p., 23).

Mind-as-computer (or at least as a computational system) is seen as being the product of a long process of adaptation in which random chance played no really significant part. "Complex adaptations . . . are so well-organized and such good engineering solutions to adaptive problems that a chance coordination between problems and solution is effectively ruled out as a plausible explanation" (Tooby and Cosmides, 1992, p, 20, 62). As this approach has developed, it has increasingly been argued that evolution is not really about reproductive behaviour as such but about *adaptation*; ". . . I believe that everything the mind does is biologically adaptive. . ." (Pinker, 1998, p., 524); and this does not mean adaptiveness or adaptation in any general sense as an on-going process; ". . .the basic experiential goals that motivate human behaviour are both inflexible and specific. . .human beings . . . have been designed by selection to strive for *specific* goals, not the *general* goal of reproduction-maximization" (Symons, 1992, pp., 138-139, his emphasis).

Not only, then, is adaptation brought back in by this approach, but so too is *design*. A claim for adaptation is ". . .a claim about design" (Symons, 1992, p., 140). "Because an evolutionary perspective suggests that there will be a close functional mesh between adaptive problems and the design features of the mechanisms one *might expect* natural selection to have produced to solve them. . ." (Cosmides, Tooby and Barkow, 1992, p., 6, see also pp., 6-9, my emphasis). "The mind, like the Apollo spacecraft, is designed to solve many engineering problems, and thus is packed with high-tech systems each contrived to overcome its own obstacles . . .[it] has to be built out of specialized parts because it has to solve specialized problems (Pinker, 1998, pp., 4, 30).

This does suggest Dr. Pangloss who ". . . could prove admirably that there is no effect without a cause…that things cannot be otherwise for, since everything is made for an end, everything is necessarily for the best end" (Voltaire, 1759, p., 4). Certainly, it does seem that rationality and design, and by implication 'purpose' and progress, are back with a vengeance. Unfortunately, therefore, the ever present danger of teleology returns. "The evolution of adaptations by natural selection is one of the few cases in which the consequences of something can be properly used to explain its existence" (Barkow, Cosmides and Tooby, 1992, p., 625). They credit Dawkins (1986) with this insight. Now, Dawkins does make statements like this from time to time (Cf.., 1986, 1997), but he certainly does not seem to be a hard-core adaptionist, and he has not reduced natural selection to this teleological basis (but see the view of Gould, 1997).

In Dawkins' work, reproductive success is still a key and it is not dependent on a perfect adaptation but rather ". . .processes which create an illusion of design." (1997, p., 4). Moreover, "…the individual body, so familiar to us on our planet, did not have to evolve. The only kind of entity that has to exist in order for like to arise anywhere in the universe is the immortal replicator" (Dawkins, 1989, p., 266).

More significantly – whatever Dawkins' exact view – there seems to be a fundamental problem with treating processes of adaptation as proximate causes (it seems very likely that many of these writers really want adaptation to be an *ultimate cause* to escape the problems of discovering proximate causes). But to the problem. How can we ever claim that something is more or less adapted?; how are we to know, for example, what *specific* task, certain morphological features or behaviours are adapted for? Take human hands, for example. Are they adapted for tool use/making, picking fruit, carrying a baby, carrying food, caressing the breasts of a potential lover, feeding self/others, throwing a spear, masturbating, doing the sign of the cross, guiding a penis into a vagina, hitting someone, signalling surrender, restraining someone, making rude gestures or sending an e-mail? Or the eye, so 'beautifully adapted to seeing'; is it adapted to seeing: predators, parents, forests, food, friends, enemies, sexually alluring features, anger, deceit, love or what?

Even if we had considerably more information than we do about sexual/social life at the point of human differentiation from a common ape ancestor, it would be very difficult to argue the primacy of any selected group of things a hand or an eye can do, let alone any single, specific one for each. And even if some of these were important then there is no reason to believe that this has not changed during the period of human evolution. To take another example; *learning* is a generalist strategy not a specific one. Who is to say that learning where the buffalo are is more or less adaptive than who is a possible mate, or who is a possible enemy, or who might cheat on one, sexually or in business dealings? It does not seem that the processes of natural and sexual selection are both ". . .inflexible and [directed only to] *specific* goals, not the *general* goal of reproduction-maximization" (Symons, 1992, pp, 138-139, his emphasis). Whatever the hand, eye or learning have done, they must not have prevented relative reproductive success; they probably aided it in a number of ways, differently at different times most likely.

What about species in general (or given individuals)?; how can we claim they are adapted? Environments continually change; so at what point is it possible to say that a species has become adapted? Even if we could agree a set of criteria for adaptation to a particular environment, by the time adaptation takes place the environment will have moved on. Moreover, sexual selection *may* move a species (or given individuals) away from adaptability to an environment; it may be harder and harder for certain individuals to live and survive in a specific environment but, nevertheless, they may be the most sought after for reproductive interests, and so increasingly outnumber those better adapted in terms of survival. This is as much a part of evolution as the evolution of the eye – supposedly a miracle of intricate adaptation. It could, of course, be argued that such sexual success is, in fact, an adaptation. Yes, but that renders the notion of adaptation tautological and rather meaningless. Even when a balance between natural and sexual selection is more or less achieved, it is unlikely that a specific balance could ever become a permanent adaptation; environments change, chance intervenes.

Indeed, what about the issue of chance? Is evolution directed or largely a product of chance?; that is, to what extent is it random? Psycho-biologists in general are intent on eliminating chance from the equation more or less completely. They accept that mutations largely occur by chance but that subsequently natural selection is a 'directing' process which gives every appearance of creating ever better designs. They are not keen on "the Law of Higgledy-Piggledy", as Sir John Herschel called Darwin's contribution to our understanding of evolution (Dawkins, 1997, p, 67). In this they have a strong ally in Richard Dawkins (1997), who argues that living things are "designoid" objects which are not the result of chance. "They have in fact been shaped by a magnificently non-random process which creates an almost perfect illusion of design. . ." (p., 4). So there is no Grand Designer but there is a design process; "Darwinism is *not* a theory of random chance. It is a theory of random mutation plus *non-random* cumulative natural selection." (p., 66, his emphasis), and ". . .because genetics enables cumulative gain the best specimen you can find in a generation is better than the best you can find in an earlier generation" (p, 21). And, thus, progress (again a hint of an ultimate cause) comes a step nearer to returning to the centre of concern

So, chance is largely rejected, luck is not worth considering and a non-rational process found to be no basis for an explanation of evolution. In terms of behaviour, maladaptive behaviour is squeezed out of any explanation. In the place of the above we find a rational universe in which design and functionalism reign supreme. But, it does seem to me (see also, Gould, 1997) that, while agreeing that evolution is not a completely random process, chance plays a much bigger part than is allowed for by many psycho-biologists and Dawkins. As noted, they generally accept that genetic mutation is random but want to stop there. But there are a number of points during the reproductive process in which chance can intervene. A random mutation can affect other genes which may have randomly accumulated over a very long time but hitherto have been inactive, unleashing quite unpredictable results. Or, mutant genes may affect embryological development in the presence of certain randomly occurring biochemical or physical events. Many genes are regulator genes with differential activation depending on circumstances and conditions. Moreover, chromosomal inversions and translocations seem to add another possibility for randomness (on all these processses Cf., Lewontin, 1974, 1978; Ayala, 1978; Harris, 1980; Washburn, 1963a; Gould, 1977; Dawkins, 1989; Ridley, 1999).

Whether some of these processes take place in an isolated group or within a very large gene pool, in relatively stable or unstable environments, in dominant or subordinate, reproductively attractive or unattractive individuals, for example, can all be a matter of 'luck' (randomness). But the effect on evolution can be considerable. The effects of natural catastrophes (Darwin not withstanding) - quite apart from the unpredictability of environments generally - certainly have an impact on evolution (Fortey, 1998; Gould, 1997). These are, as far as we can make out, random events – at least as far as organic evolution is concerned.

It might be unfair to do so, but it seems to me possible to use much of Dawkins' own work (Cf., 1976, 1989, 1997) to argue that at times he has overstated the significance of adaptation at the expense of chance. For example, he accepts that environments are probably too unpredictable for mutations to become 'directed'; that such things as stored up mutations, 'segregation distorter genes' and the extensive existence of 'polygenes' can all have effects (regardless of their adaptive value) on embryological development - greatly affecting phenotypical outcome (regardless of adaptive value); that some embryological processes may be better at evolving than others; that different 'evolutionary stable strategies' (ESSs) may result in 'the best of a bad job' reproductive practices in which the best (presumably best adapted) do not always have the greatest reproductive success; that if failure is not too costly there is no selection pressure to improve; that, according to the 'handicap principle' (Zahavi, 1975), in some cases the positively handicapped have very good reproductive success; that the race between eater and eaten (including at the level of parasite and host) is an ongoing struggle to stay even and success can only be measured against each other not necessarily in any 'objective' terms of adaptation (now or at some past time).

He accepts that limited 'punctuated equilibrium' can exist, opening the door to a certain number of random events in the history of a species. His descriptions of complex ESSs are extremely perceptive – if not brilliant - but suggest the vulnerability of the evolutionary process to random events other than just random mutations. There are, in his formulation, also the possibilities of stored up genes being randomly activated and of embryological scenarios and geological and physical events having unforeseeable causes and effects. Thus we have the possibility of an evolutionary flow which is not moving in an adaptive direction (by anyone's non-tautological definition of adaptation), but which will likely continue because the evolutionary cost (ontological disruption) of mutations which might turn it in another direction would be too great.

At one point while arguing for the non-random nature of natural and sexual selection, he speaks of a "gene sieve" through which each generation passes and only an "elite" emerges. The elite

> ". . .have what it takes to get through sieves. They have [created] a million bodies without a single failure [each of which] has survived to adulthood. Not one of them was too unattractive to find a mate. . .Every single one of them proved capable of bearing or begetting at least one child " (Dawkins, 1997, p., 76).

Well, this is not exactly my idea of an elite. It sounds more like the lucky, if not the just-scraping-by end of the struggle for reproductive success. The only definition of non-failure (and by implication, adaptation) is that they survived to enough of adulthood to pass on genes. It is true that they ". . . survived ice ages and droughts, plagues and predators, busts and booms of population . . . " (p, 77) and so on, but this introduces even more possibilities for the intervention of chance into the evolutionary process. Who knows what genes were sitting next to the volcano when it erupted, or who, engrossed in fantasy when out for a walk, stumbled into the path of a hungry predator, or who happened to live far from the river when all the lakes dried up, or was the handsomest of all until frost bite got his nose (and rumour had it one or two other things as well), and so on and on?

Such ambiguity in the evolutionary process does seem to be very much in line with Darwin's own observations in the *Origins* (1859 [1996], esp, Chaps. 3 and 4). Despite his rejection of 'catastrophism' as an explanation of evolution, and his occasional use of terms such as "exquisite adaptations" and of chance being a 'false view', and his comment that in nature it is clear ". . . how infinitely complex and close-fitting are the mutual relations of all organic beings to each other and to their physical conditions of life" (p., 51, 67), it seems, nevertheless, that Darwin was talking about a process in which chance and luck can intervene at a multitude of points while 'achieving and maintaining' this state. His description of 1000 seeds resulting in only one success, depending on an extremely complex interaction of the state of the mistletoe and apples and parasites and birds; or his constant references to "checks" on inevitable population increases as being, in his vivid examples, combinations of the availability of food, the existence of predators, climate fluctuations, insects, changed vegetation, cats, field mice, bees, nearby villages, heartsease, red clover - all extraordinary interactions - represent, not only "some of his best writing" (Ridley, 1994, p., 86), but, it seems to me, a clear recognition that evolution is not a slick, rational process leading to perfect adaptation, but rather one subject to a multitude of variabilities.

Such an interpretation is reinforced by numerous of his own direct comments. He points out that the notion of a "struggle for existence" is used as a term of convenience not a law of nature; that the success of a number of individuals of a given species is often because ". . . the conditions of life have been favourable and that there has consequently been less destruction of the old and young, and that nearly all of the young have been able to breed" (pp, 53, 55). So, external events (often random to the evolutionary process) have an enormous effect. ". . . highly favourable circumstances . . . checks. . .relations. .. are complex. . ." and often "unexpected" (*Origins*, pp., 58-59). A single event can have far reaching, unexpected effects. He gives a complex example of the introduction of a new species of fir tree to a heath:

"Here we see how potent has been the effect of the introduction of a single tree [species]" (p., 60); but immediately enclosures of the common lands becomes important because otherwise cattle would have prevented the fir trees growing; but, of course, cattle are dependent on insects. In the end

> "No country can be named in which all the native inhabitants are now so perfectly adapted to each other and to the physical conditions under which they live, that none of them could be improved. . .[so we need not be surprised if] . . .inhabitants of any one country, although on the ordinary view supposed to have been specially created and adapted for that country, being beaten and supplanted by the naturalized productions from another land. Nor ought we to marvel if all the contrivances in nature be not, as far as we can judge, absolutely perfect . . ." (*Origins,* p, 69).

Although elsewhere he mused that "the wonder indeed is on the theory of natural selection, that more cases of the want of absolute perfection have not been observed" (p, 381), nevertheless, it remains clear that for him even the *processes of adaptation* were subordinate to the issue of reproduction. So important was this that one has to be careful not to interpret Darwin's account of natural selection as being a conscious process; all living things are "striving to increase at a geometrical ratio", he continually emphasized (e.g., p., 65), but, of course not all (often not even most) can; thus we have natural selection which is not conscious. Very telling, it seems to me, is that while ". . .adaptive characteristics, though of paramount importance to the being, are of hardly any importance in classification." (p., 386). In other words, species are not identifiable through adaptation but instead through their ability to leave copies of themselves, that is, a consistency of type.

Thus, the selfish gene, which, to my mind, is a powerful notion, comes into its own. It means that those genes that have most successfully reproduced themselves are what we see; it means that we need to look at reproductive processes and for the capacities and motivators which underpin them. It does not mean, however, that genes are selfish in a conscious way. Genes are not trying to succeed; they do not have a will; they are not compelled by some force of nature to be rational in this activity. Dawkins and psychobiologists, of course, make this point over and over again, but insist that it does not mean that we are reduced to the "the Law of Higgledy-Piggledy". No, we are not subject to the Law of Higgledy-Piggledy because natural selection is not a random process (they argue), it is a 'rational' process through which progress is made through ever perfecting adaptations. So natural selection takes over from God, the Grand Designer, as the design force. It is, however, secular, it is 'scientific'. Moreover, it is not about sex, jealousy, envy, love, hate, deceit, a will to power; it is about algorithmic, rational, mathematical, mind-as-computer guided genes; genes inevitably headed for more and more perfect adaptations.

Unfortunately, to make adaptation, design and functionality key to an explanation inevitably invites teleology, implying that a greater rational design exists – God's or Nature's, no matter. To say that natural selection is a special means of achieving design, adaptation or perfection is not much of an advance over saying that it is God's Will that these be achieved. A great deal of the immense contribution of Darwin, and the more recent creative biology of people such as J. B. S. Haldane, W. D. Hamilton, George Williams, E. O. Wilson, Richard Alexander, Stephen J. Gould, Richard Dawkins and Robert Trivers, for example, are lost in 'mind-as-computer'/a process of perfect adaptation (presumably for the eventual better understanding and 'progress' of humans).

Indeed, it is not difficult to imagine that within the 'design/adaptation perspective' a search for purpose, progress and directionality lurk in the minds of certain psychobiologists. They may or may not deny this, but certainly would argue that they are not teleological because, they would claim, natural selection is separate from its outcomes; the appearance of design is an inevitable product of natural selection rather than of a Grand Design. Well OK. This is certainly better than a theological Grand Design approach. But I am not sure that it is better than social science functionalism. Spencer and Parsons, for example, would have argued more or less the same thing. They would have said that human *rationality* is separate from its outcomes and that there probably is no theological Grand Design. Nevertheless, they would have pointed to human capacities for instrumental rationality and foresight 'leading' to societies as being functional wholes (the whole being greater than the sum of it parts) as evidence of rational design at work. Moreover, they would have argued that because learning is cumulative, history and societies are progressive.

The only reason I can fathom for this development among certain biologists (specifically certain psychobiologists) is that they are extremely desperate to avoid the Higgledy-Piggledy label (which carries with it the notion that theirs is not such a hard headed, absolute science after all; that their science contains some messy bits), and/or to distance themselves from the social sciences (see Chapter Two), which they tended to lump, in some cases, with 'creation science' (Cf., Symons, 1992). Indeed, the more social scientists became interested in translating the notion of reproductive success into some kinds of motivators of human behaviour, the more some psychobiologists seemed to fear pollution of their quest to get back to the scientific and moral high ground, as they see it, of ultimate causes (thereby avoiding the muddy waters of such things as lust, jealousy, envy, hatred, fear or a will to power, for example).

Among psychobiologists (he prefers the term, 'evolutionary psychologist'). Steven Pinker (1998) does allow some room for psychological concepts such as these. Despite his search for evolutionary adaptation, he recognizes that all things are not necessarily perfectly adapted (especially some time after they originally evolved), and that emotions play an essential part in motivating behaviour. Nevertheless, while emotions can cause humans a great deal of anguish, and be very disruptive for groups, they are adaptations in so far as they *necessarily* provide reproductive success. ". . . emotions are adaptations, well-engineered software modules that work in harmony with the intellect and are indispensable to the functioning of the whole mind" (p., 370). Emotions set the highest level goals for the brain, and ". . .trigger. . .the cascade of subgoals and sub-subgoals that we call thinking and acting." (p., 373).

So, emotions too are part of the rational calculus of cost-benefit analysis of the selfish-gene. We are told that emotions such as liking, anger, gratitude, sympathy, guilt and shame make sure we get the right friends and mates. Well this may be true in the long evolutionary run of things for humans generally - if by 'right' we mean those 'friends' who have been party to our specific combination of species typical genes still being in the gene pool - but it can be very misleading if we are looking to emotions as *proximate causes of individual and general social behaviour*. We do not always love those who are nice to us (or faithful to us); nor do we always befriend those who best serve our reproductive interests. And we do not just feel guilty when an unconscious cost-benefit analysis of our reproductive interests calls for it – with guilt we often blame *ourselves* for transgressions against ourselves *by others* (and we often retreat from a number of potential reproductive opportunities). We do not just show gratitude because if we do it might be reciprocated; we do not have a pecking order because nature is hierarchical and it is *efficient* to have one – we do so because a whole assortment of emotions motivate us to continually evaluate social others, seek status and to distance ourselves from social others. Depression is powerful – it causes all sorts of behaviours which are difficult to understand in terms of mathematical calculations. The human will to power manifests itself in a whole variety of socially and politically significant ways – but interestingly is never mentioned by sociobiologists and psychobiologists.

There is no doubt the above mentioned emotions and behaviours, however they are manifest, all of them, have had an evolutionary origin (but not necessarily as part of a perfecting design process). But, accepting an evolutionary origin of emotions (and desires and fears) in and of itself, is not an explanation of their *consequences*, as causes of individual behaviour and of general social patterns in modern society. And it is an understanding of *these* consequences which we will need for a non-teleological social science. Human social behaviour is a product of human nature, a human nature created by natural selection in a different phylogenetic context, but it nevertheless remains a bundle of causes in the present – causes which are not easily explained in terms of cost-benefit analysis. We will have to discover, for example, how a male infant's lust for a mother's embrace and, at the same time, a will to independence, affects their psycho-social development - including development of a sexual identity - regardless of why these evolved in the first place. To what extent does this lead males to search for mates who are both mothers and whores, and what kinds of taboos, rules and moralities do we invent to cope with this?

How are lust and love and jealousy and separation anxiety, for example, balanced into different kinds of mating patterns? We will want to discover how different types of guilt and envy and empathy and a fear of loss, for example, motivate identifiable propensities for tabooizing, binding, distancing and social bounding, for example, in specific situations - and then, what forms, or patterns, of behaviour might emerge from these (in various contexts)? We will have to identify patterns of emotional expressions, desires and fears which motivate and underpin the politics of vengeance seeking through, variously: blood feuding, forming alliances, demanding compensation and arguing for laws, for example; what about the emotions, desires and fears which underpin social networks of live and let live 'social worlds', hierarchically graded and politically administered? How can we explain the ubiquity of politics in all human social relations (with a small p at least) – with all the deceits, self deceits, obsessions, angers, hates and dedication these so often entail? Why do humans build pyramids? Why are we seemingly never completely satisfied (for very long at least)?; why is happiness so elusive? And, most significantly, how do these factors affect social change (human history)?

Once we have accepted the importance of an evolutionary perspective in discovering *what* emotions, desires, fears and patterns of cognition *exist*, and that they are extremely important as motivators of behaviour, we must push our causal analysis forward, not turn the trajectory into a circle, driving it back to teleology. Unfortunately it is in the forward trajectory that psychobiologists have resisted. They, in fact, quickly jump back to a search for an *ultimate* cause (from among: replication, reproductive success, adaptation, design) in order to be able to claim a right to explain social behaviour without getting their hands dirty with messy concepts. Personal and social ". . . battles are driven by strong emotions, but emotions are not an alternative to the genetic analysis; the analysis explains why they exist" (Pinker, 1998, p., 458). And remember, the genetic analysis – the why - tells us that emotions are 'rational' cost-benefit calculators with only the selfish gene's interests at heart. "To understand cooperation and conflict, we have to look to the mathematics of games and to economic modeling" (Pinker, 1998, p., 40) because social psychology seems ". . .to fall out of a few *assumptions* about kin selection, parental investment, reciprocal altruism and the computational theory of mind." (Pinker, 1998, p., 517, my emphasis).

We are back to descriptive concepts of social behaviour as teleological explanations of the selfish gene. As a result we get statements like, "The love of kin comes naturally; the love of non kin does not. . . .*Homo sapiens* is obsessed with kinship . . . [foraging people] . . . rattle off endless genealogies. . .in all cultures marriages are alliances between clans, not just spouses. . .dowries and bride-prices are ubiquitous in human cultures. . ." (Pinker, 1989, pp., 429-437). Virtually no evidence is given for these statements and, in fact, none stand up to empirical evidence (see note two, p, 234). But kin selection suggests adaptation, design, a moral universe, *despite the selfish gene*; it suggests rational humans – so in the end even emotions such as jealousy or lust are designed to get us there – wherever *there* is. Teleology is mighty difficult to escape.

It has been a central argument of this work that a major contribution of Darwin was to bring the possibility of chance and non-purpose into our explanations; that design as such was to be rejected. It has been my contention that we must be willing to accept the whole brunt (all implications) of this challenge in order to escape teleology in the social sciences. From the realist's consideration of human history, from Darwin's observations and from a consideration of modern history (not to speak of quantum physics), there is every reason to believe that the full de-centering of humankind by Darwin was on the right track. One does not have to have God-like perception to realize that neither history nor modern social life is a picture of mind-as-computer, perfectly adapted to environments in terms of survival efficiency, pain free repose, perpetual happiness or social tranquillity, or whatever else might be used as a measure of adaptation.

Darwin's great contribution was to escape teleology. Not only must social sciences do the same, but modern biology must avoid slipping back into it. Evolution is about adapt*ing* and *reproductive success*, not 'Adaptation'. It is virtually impossible to argue from the perspective of functionality or adaptations without being circular, tautological or teleological; any illusion that it gives of being more scientific is just that, an illusion. Besides teleology, this line of reasoning avoids the real problem of finding a believable link between biology and behaviour. In it, we are asked to jump from genes as *causes* of behaviour, operating through some mechanisms of mind, to the epistemological issues of 'what is the nature of evolution?', and then on to the ontological issue of 'what is the nature of evolution's final great adaptation?', as an explanation of human existence. Rather then seek proximate causes of behaviour, this approach leapfrogs to an epistemological argument (claims about the 'true' nature of evolution), which is then treated as an *ultimate cause* (evolution as *Adaptation*) so that the jump to the descriptive category of mind-as-computer seems to make sense. Rationalism and human ethnocentrism die hard.

There are, it seems, at least two crucial problems involved here. One is that sociobiologists (and psychobiologists even more so) have hesitated to invent or utilize what might appear as non-rationalistic concepts, either because they have not thought it necessary or because they are over-eager to show the relevance of their work in terms of existing concepts. The second is that sociobiologists and psychobiologists have been too uncritical in their analysis of the problems of causation and have too quickly latched onto the descriptive concepts of behaviour discussed above (e.g., altruism), or the notion of mind-as-computer as, for all practical purposes, explanations (causes) of human social behaviour and human cultural patterns.

However, if their findings are to revolutionize the human sciences, as they often claim (Cf., Wilson, 1975; Alexander, 1979a; Tooby and Cosmides, 1992; Pinker, 1998; Blackmore, 1999), it seems only natural that sociobiologists and psychobiologists should come up with some new revolutionary concepts, or at least it might be better for understanding if they do so. More parsimoniously, they could use some concepts adapted from sexology, psychobiology, primatology and paleoanthropology to aid them in applying principles of population genetics and sexual selection to study the evolution of possible proximate mechanisms, especially of possible emotional motivators and patterns of general cognition. This would require, among other things, theorizing about possible reproductive competitions and ecological conditions during the period of human species differentiation.

Social Scientists

If straight genetic determinism, descriptive concepts and 'mind-as-computer/the ultimate evolutionary adaptation', are not the answers to the problem of proximate mechanisms, what about learning, reason and social constructionism, the favourite proximate mechanisms of social scientists? Some social scientists have gone so far as to suggest that because humans have such an infinite capacity to learn, and to reason and invent, it should be obvious that biology generally, let alone genes, has no determining power over significant human social and cultural behaviour. Further, they point to what seems to be a great variety of behaviours and cultures found around the world as evidence of this. Therefore, they conclude, sociobiologists must surely be wrong (Cf, Harris, 1979; Leach, 1982; Sahlins, 1976).

Yet has there been an infinite variety of behavioural and cultural patterns during human history? If so, would political science, social economics, anthropology or sociology as abstracting disciplines have been possible? I think the answer is a clear 'no' to both of these questions (see also Trigg, 1982). As an example take kinship, a hotly contested area between the two camps (see esp. Sahlins, 1976 for social science; Alexander, 1979 for sociobiology). Van den Berghe`s (1979) work illustrates, in fact, that a very limited number of kinship patterns exist, although many more can be conceived of by the human mind. Social scientists themselves would, I think, have to agree that there is, in fact, a very limited number of patterns which social hierarchies and political 'systems' have taken during the course of human existence, even though, again, the human mind can think up many more. For example, where is the society in which those with lowest status are in positions of political authority?; or where is the society in which teenagers rule, or slaves have the highest material standard of living, and so forth?

Some social scientists consider that they have solved this problem through the 'discovery' of synchronic, self-regulating, ('rational') patterns of culture and/or social structures. These patterns are held to have greatly reduced the possibilities of unlimited variations historically but do not preclude the development of a reformed future. Although socialization (the processes of culture transmission and social reproduction through time) is, they usually argue, conservative, in theory it progressively gives way to reason and thus to a closer link with the larger 'force' of universal rationality.

Unfortunately, the postulated synchronic cultural and/or social patterns described here have not been based upon studies of physiological, or even psychological, propensities, often not even on history or anthropology (Somit, 1981), but instead on the teleologically derived, holistic needs of social, economic or political 'structures' as abstractions. Most often these, in turn, are based only on particular desires of what humans and human societies ought to be like. In extreme cases 'systems' or 'structures' are deified to the point where they are said to have their own powers of causation with regard to, if not all aspects of individual human behaviour, certainly to aggregate behaviour, and to the replication of the system or structure itself almost independently of human action.

So, in some circles, such as among British social anthropologists, Durkheimian and functionalist sociologists, Marxists and structuralists generally, society and/or systems became reified to the point that their existences were taken as almost mystically given. But this leaves the difficult question of explaining how these systems got there in the first place. This position involves treating an abstraction, society or 'system' as causing observable behaviour. This is a version of ultimate cause rather than a real attempt to deal with proximate cause; one, however, in which we might as well treat God as the ultimate cause since, as a concept, He/She/It is no more abstract than the concept of society.

The above is the social constructionist, or social deterministic, view in its most crude, or some might say abstract, form. In its basic, less rarefied, form, social constructionism argues that 'meaning' in human social relations is said to be determined by the nature of specific social interactions. This implies a multitude of possible meanings depending on a multitude of circumstances. But a consequence of permitting this degree of flexibility almost precludes the possibility of postulating more stable, regular patterns of behaviour. That is, society becomes difficult to conceptualize even though at one level it is being treated as the Final Cause.

The social constructionist notion is paradoxical in another sense as well. In most monkey species and among herd and pack species generally, for example, there is what appears to be a very powerful social determination of behaviour. That is, most of the instincts, drives, 'emotions' and patterns of easy learning in such species make it very difficult for individual members to act independently of their group, sometimes even 'preventing' the protecting of their own offspring if it means leaving the proximity of their own group.

In this sense, social determination in nature is in direct conflict with reason or learning. It is, in fact, one of the clearest examples of genetic determinism in higher mammals. Yet social constructionism is often seen by social scientists as being a means of declaring humans free from the iron determinism of their biology. This might be because social scientists have never really investigated what it might mean when they claim that 'such and such' is not 'natural' but rather is socially constructed. For, it does seem that, during their evolution, humans actually became more free of other members of their own species, indeed, more free from members of their own local groups, than is true in most primate species. There was an evolution away from genes working through social determination, not one towards it.

Proximate mechanisms for a majority of social scientists, then, are: 'rationally' driven, learned behaviours, directed toward constructing functionally coherent market behaviours, 'economic systems', social institutions, social structures/systems and/or cultural patterns. Sometimes this requires first recognizing major imperfections (dysfunctions) of given systems and participating in 'social constructing' through either reforming or overthrowing them. These mechanisms are perceived to be taking us on the way, progressively, to human perfectibility, if not to social utopias.

This form of teleology is the process of myth making, if not religious enthusiasm, and it is little wonder that sociobiologists and psychobiologists fret about the state of the social sciences. But, as we have seen, the proximate mechanisms of sociobiologists are more descriptive than causal. In the end we need more than descriptions and teleologies, and the simple tautology of 'inclusive fitness' and/or claims that 'rationality' is the driving force of human activity will not do in their stead. Still, the problem of causation is crucial. Indeed, it is my argument that to advance the study of human behaviour we must discover species typical motivators (proximate causes) of it. The problem is to do so non-teleologically and non-tautologically. Before we can suggest possible ways forward, then, we must agree some concept of causation.

Causation and Human Nature

I think we can learn from Rosenberg (1981; see also, Salmon, 1984; Bhaskar, 1986; Rescher, 1987) in this matter. He argues that to have an empirical science which includes plausible claims about causes we must locate "natural kinds". By this he means units which can stand in a causal relationship to each other, but be identified, measured and conceptualized separately. By causal relationships he means ". . . all contingent connections . . ." which can denote determination in the sense that they represent observable, or at least postulated, mechanisms and regular procedures by which natural kinds can be said to affect one another (p., 52). For example, bio-chemists can identify, study, measure and theorize about genes separately from the morphological results of those genes. Morphologies in turn can be identified, studied, measured and theorized about separately from genes. Nevertheless, general statements can be made about contingent connections between them (conceptual propositions in the field of embryology).

It is important to note that this does not necessarily mean that we can discover a pre-determined path or direction of 'natural' evolution. Determination in the above sense means nothing more than that deterministic statements in the form of contingent connections can be made. Prediction requires that we know every possible combination of variables and the contingent connections between them, now and in the future. It is highly unlikely that this can be achieved in the social sciences because humans to some extent change both the nature of the variables and the connections between them. Indeed, this has yet to be achieved in the physical sciences. The neo-Darwinian theory of evolution, for example, provides contingent connections between environments, genes, reproductive behaviour and phylogenesis but in no way attempts to predict the future evolution of species.

The discovery of a number of contingent connections nevertheless should lead to probabilistic (law-like) statements which would enable us to say that when certain conditions are present, other things being equal, such and such is likely to follow. But other things are rarely equal, and so the search for causes in the form of contingent connections is a never-ending process (Rosenberg, 1981). From such a search, however, we can perhaps continually improve our understanding of the relationship between events and processes in the present. That is, although we may not be able to predict the direction of organic and social evolution we can, by uncovering recurring conjunctions, make the present more comprehensible than it was for previous generations, and that can contain lessons for future generations.

Rosenberg's approach implies a need to be able to conceptually 'reduce' any one level of analysis to the one below it and to be able to conceptually 'expand' a given level to the one above it - that is, to be able to make contingent connections between levels (see also, Brown, 2000). This brings him into direct conflict with most of the social sciences, especially sociology, where reductionism is considered to be a sin of enormous proportions (Herbert, 1983). In sociology a sure way of instantly dismissing certain scientific points of view is to accuse their proponents of being reductionists. This is based on the Durkheimian rule that 'social facts' are to be explained by other social facts and never by psychological facts. Although this rule was established partially in order to give sociology its independence from psychology (Rosenberg, 1981), and to conceptually free humans from control by base, animal emotions, it nevertheless has some merit. It is unacceptable reductionism to say, for example, that modern societies are products of genes for both selfishness and altruism, or that civilization is a product of repressed sexuality.

These are examples of unacceptable reductionism because they do not show us the proximate causes of altruism, selfishness or repression. So many steps have been left out that the causal explanations look ridiculous to anyone not expert in the thinking that led to the statements. Moreover, there is no guarantee that the steps can be made. I doubt that they can for altruism and selfishness, since these are descriptive concepts and probably will not derive from the 'natural kind' of genes. Sexual repression and civilization is another matter, but it will take many steps before contingent connections can be established. Nevertheless, along with Rosenberg, I would thus argue that social facts must be reducible to psychological facts or they are not facts worth having (see also, Buss, 1994; Dennett, 1995; Pinker, 1998; Badcock, 1986, 1994). That is, the best (and sometimes only) test we have for many of our scientific concepts is to see if they can be linked with concepts being developed at another level. But this must be done step by step, with testable - or at least reasonable, given a certain stage of knowledge - concepts. If we cannot make contingent connections between levels, one or both sets of concepts are in drastic need of revision and even perhaps demolition. At the same time we cannot make reductions based on gigantic academic leaps, analogies or homologies, and expect them to be taken as serious causal statements.

Rosenberg does not seem to think that there is a natural kind linking genes and behaviour. In this book I will argue that there is; that human nature (as an interaction of cognition and emotional processes) can be conceptualized separately from genes and from behaviour, but nevertheless that contingent connections can be made in both directions. It is because they have ignored human nature that neither biologists nor social scientists have been able to make believable contingent connections; there are few, or none, between genes and behaviour, but they exist between genes and human nature, and between human nature and behaviour.

Human nature as a natural kind between genes and behaviour cannot be seen as simply some sort of machine for translating genetic messages into behaviour. Nor can it be seen as the seat of mystical humanness, as some version of a soul. Rather, human nature must be conceptualized as bio-electric neural patterns and processes with the capacity for emotional and cognitive outcomes (see Chapters Four, Five and Six). These outcomes must be shown to be vulnerable to influences by environmental conditions, including the biochemical environment in which genes instruct for their development. Outcomes are to be treated as motivators of behaviour and of learning. Motivators operate through consciously controlled feelings, desires and fears, and also through much less controlled emotions and feelings. All together this package results in both relatively pre-determined and variable outcomes. Human nature, in other words, is the interface between the permanency of genetic replication and the human capacity to feel a range of, not always consistent, emotions and to learn a variety of responses to environmental stimuli.

The problem of permanency, however, raises another area of disagreement between biologists and social scientists. Some sociobiologists and psychobiologists would argue that genes provide consistency and that looking for something called human nature is a typical social science waste of time. Many social scientists would argue that such a view is genetic deter-minism and reductionism of the worst kind, and, furthermore, that there is no such thing as human nature; human behaviour, they would argue, is infi-nitely variable. As noted, I will argue that it is possible to meet in the middle with a conceptualization of human nature which goes some way to satisfying both sides. But to maintain the degree of consistency which is evidenced by the existence of the species *Homo sapiens*, our concept of human nature must contain something like *'pre-knowledge'* . Such a notion suggests that we learn much of our behaviour but that we are more or less programmed to learn some things, *in the right conditions*, and that some things will be almost impossible to learn in any conditions. Thus we have a formula for consistency along with variability (and the possibility of searching for what is consistent and what is variable).

It is very important to develop an understanding of the working processes of pre-knowledge because without such knowledge we are left with genetic determinism in a most robot-like fashion, on the one hand, or no hope for a science of humans, on the other. Yet, despite numerous attempts, by both biologists and social scientists, little progress has been made in achieving an agreed 'middle', let alone one that works. As with the issue of cause discussed above, this has had as much to do with the different histories of the two disciplinary areas as any intrinsic epistemological or ontological barriers between biology and the social sciences or between humans and other animals. It is almost as if both sides want to stay away from this dangerous area, preferring instead to try to leap over it from the safety of their own discipline. But if the gap is to be bridged, some of these disagreements will have to come out into the open.

HUMAN NATURE AND THE PROBLEM OF 'PRE-KNOWLEDGE'

Genes are protein structures which contain both the information and ability to replicate themselves, given a sufficient supply of additional building material and energy. As such, genes represent conservative forces which resist variation and thus work against the universal tendency towards entropy (Campbell, 1984). However, learning as a concept implies the ability to absorb, interpret and remember external stimuli. Learning, by itself, encourages variation and makes no special provision against entropy. Nevertheless, learning in a controlled manner can, through the flexibility of response it provides for, make it easier for genes to survive and replicate themselves in varying environmental conditions.

Therefore, successful species (interbreeding groups which maintain, through time, similar characteristics) generally possess genes for both learning abilities and also for restraining, motivating, controlling and directing learning. In order to understand the proximate causes of species typical (relatively consistent) behaviour, biologists need to discover both the range of specific learning abilities and the pre-learned motivators and repressors inherent in that species. (Specific individual behaviour, of course, must also be derived from a consideration of these ranges interacting with specific environmental conditions.)

In the past, the concept of instincts has been used to describe pre-learned motivators for species other than humans, but instincts imply a pre-determined fixity which has seemed to be contradicted by the evidence of animal behaviour and so biologists have largely abandoned this concept (Thorpe, 1979; Goldsmith, 1991). The concept of imprinting has gone some way to bridge the gap between pre-determined behaviour and learning, and seems to fit a number of species, but even this appears to be too rigid when considering higher mammals, especially humans (Hinde, 1982). Freud postulated libidinal 'drives' to set in tandem with a 'death wish'; from the interaction between them during specified stages of psycho-sexual development, in a context of external repressors, he postulated a number of universal learned human behavioural responses (Cf., 1971, esp. Lecs. XXXI and XXXII and 1957). Few, including many psychoanalysts, have been completely satisfied with his postulations, especially concerning social behaviour. Whether this approach failed because it was never really tested or because it is fundamentally flawed remains an interesting question.

Whatever the case, humanists generally, and social scientists particularly, were looking for a much more 'positive' conception of humankind than Freud was offering (Ward, 1897, 1906; Passmore, 1970; Russett, 1976; Nisbet, 1969). Notions such as 'the psychic unity of mankind', from anthropology, and the "collective unconscious" (Jung, 1972, 1984) were attempts to provide this but they proved to be too vague to be of much use or, more likely, social scientists had lost interest in human nature by the time these notions were being developed. As the Twentieth Century study of humans unfolded, culturology, or the study of the "superorganic" (Kroeber, 1917, 1952; Boas, 1965, 1966; Mead, 1943; Benedict, 1946; White, 1949), had effectively driven biology and any type of Freudian or even semi-deterministic psychology from the field (Harris, 1968; Kuper, 1983; Freeman, 1984). Culture, as an abstraction, became the cause of human behaviour.

The Superorganic

The idea of culture fits in nicely with the desire among humanists and social scientists to distance humans from biology, a desire which was gaining momentum at the turn of the 20th Century. The concept of culture was based on mind, not matter; it was allied to reason and logic, not racial/ genetic determinism. It was a product of the human mind (not brain) working with abstract concepts and symbols in an extremely creative way. Humans were the creators of art and religion, philosophy and science, poetry, just as much as they were sexual creatures, or aggressive or anxious, or repressed. And when cultures found the best way to organize individual relationships and social groupings these last would disappear and the first bloom. Humans were perfectible through mind, and happiness was only a matter of time. The South Seas were full of societies which had achieved just this, or so it was claimed.

While the tautological and circular (somewhat mystical) nature of this approach soon made it suspect among many social scientists (see below), the idea that there is a deterministic force greater than biology, or indeed, individual will, does not easily go away. Durkheim's sociology included a notion of a 'collective conscious' as the basis of the social glue which held societies together; Parson's functionalism gave a prominent place to norms, values and beliefs as governing and guiding forces; Marxists looked at the hegemonic power of both agencies of civil society but also 'false consciousness' to explain why the revolution seemed to be somewhat delayed.

Interestingly, in very recent times, after arguing a vigorous, and very persuasive, case for the importance of utilizing the Darwinian principles of natural and sexual selection in understanding human behaviour, and for the significance of 'the selfish gene' notion, the philosopher Daniel Dennett

(1995, Chap. 12) still appears to relinquish causes of human behaviour to culture. For him, "We are different. We are the only species that has that *extra* medium of design preservation and design communication: culture" (p., 338, his emphasis). He argues that culture powerfully represses genetic processes. We have ". . .an entirely different outlook on life from any other species" (339) because culture is ". . .radically different from the way of life of all other living things" (341), with ". . .revolutionary powers" (342).

He utilizes Richard Dawkins' (1976, 1982, 1986) suggested concept of "memes" to try to establish units ". . . of cultural evolution . . ." (342) which transform the brain ". . . into something much more powerful . . ." (343) than its genetically designed hardware. At this point we have jumped to culture more or less completely because when we look at memes they are not links between genes and behaviour but are cultural artefacts, mostly ideas, pure and simple. Memes include: the arch, wheel, clothing, The *Odyssey,* de-constructionism, Greensleeves, the first four notes of Beethoven's Fifth Symphony, ideas, catch-phrases, Plato's *Republic*, Faith, bifocals, education, 'ought', good , truth, beauty, the SALT agreements. Now, there might be *pre knowledge propensities* which push us to all these things, circumstances being favourable, but these artefacts are *outcomes*, not pre-knowledge 'de-signers' or motivators.

And, Dennett very specifically rejects any linkage of memes to brain structures or processes, giving them complete independence. In fact, it is "The invasion of human brains by culture, in the form of memes [which] has created human minds . . ." (p., 369). In a recent work, Susan Blackmore (1999) has attempted to further this thinking, and, in the process, to give "Memetics" its own theoretical basis for ". . . the grand new unifying theory we need to understand human nature" (p., 9); or, in the words of Richard Dawkins in the Forward, "Any theory deserves to be given its best shot and that is what Susan Blackmore has given the theory of the meme" (p., xvi).

Unfortunately, the effect of Dennett and Blackmore seems to be to reinforce the Cartesian dualism long held to in the social sciences, but now from a purported biological direction. And this is despite specifically rejecting both the existence of such dualism or the value of explanations based on it. There is agreement that memes have already, or are about to, escape the leash of genes. Blackmore is relatively clear that 'adaptation' and reproductive advantage are no bases for an explanation of modern human behaviour. A meme is "a unit of imitation" (p, 5); ". . .we have to think of them as autonomous, selfish memes, working only to get themselves copied" (p., 8); they have "powers" and " 'interests of their own' " (p., 17). In fact, any comparison with genes must be for the purpose of analogy only.

> "This comes to the heart of the issue. For me, as for Dawkins and Dennett, memetic evolution means that people *are* different. Their ability to imitate

> creates a second replicator that acts in its own interests and can produce behaviour that is memetically adaptive but biologically maladaptive" (p., 35, her emphasis).

Memes have become the "tools with which we think.. . . [they] . . .provide the driving force behind what we do and the tools with which we do it" (p., 15 , 171), but, as with Dennett, memes are not dependent on the physiology or psychology of the human brain. They are, moreover, for Blackmore, not dependent on human consciousness, or at least self-consciousness.

> "The answer is to have faith in the Memetic view; to accept that the selection of genes and memes will determine the action and there is no need for an extra 'Me' to get involved. To live honestly, I must just get out of the way and allow decisions to make themselves. . . Clever thinking brains, installed with plenty of memes, are quite capable of making sound decisions without a selfplex messing them up. . . Replicating power is the only design process we know that can do the job. . . [of creative production]" (pp., 244-245, 240).

So what are the characteristics of this all-powerful, independent force which has risen above genes? Above all, it is the capacity to be a replicator based on the human ability to imitate. The analogy with Darwinism is that the process is mindless but nevertheless follows "Dennett's evolutionary algorithm" (p., 14) whereby more memes are created than can survive and that there is a process of selection in what ends up being retained, indeed, in what will thrive, implying an increasing improvement of (better adapted) type. Genes are a great replicator, but the argument is that with human evolution a second replicator has been unleashed, a replicator with its own interests and evolutionary dynamic. At first genes were selected on the basis of genetic advantage; So "we like cool drinks and sweet foods, and enjoy sex . . ." (p., 78), but, it is postulated, this process would have inevitably evolved to a point ". . .which would allow the memes to outwit the genes" (p., 78).

Genetic selection is tricked into selecting for memes with the greatest capacity to copy themselves, regardless of what behaviour they may otherwise cause. And like runaway sexual selection, where big, colourful tails and 'sexy sons' might have a sexual (thus reproductive) advantage, for a time at least, memetic selection comes to favour the best copier regardless of what is being copied; but unlike runaway sexual selection which is eventually leashed in by genetic selection, memetic selection escapes completely. So, some people do not have children ". . . not by conscious design or foresight, but simply because they [memes for not having children] are replicators" (p., 142). Memes create religions, and, as a product, we can observe that: "The history of warfare is largely a history of people killing each other for religious reasons" (p., 199).

Memes often seem to have a 'conscious will' with the ability to control their own destinies and their evolution is definitely progressive. They can jump from one mind to another, from mind to book, from book to mind, from mind to computer, and so on. They have restructured, if not created, the human brain, created language, writers, journalists, broadcasters, film stars and musicians and computers - a product not of designed conscious control but of ". . . the playing out of the evolutionary algorithm" (p., 204). Language evolved as a product of the evolution of memes and is one of the chief means of their replication (and dissemination); "Roads, railways, and airlines connect larger and larger numbers of people together, just as common languages and writing systems do" (pp, 210-211), all to ensure the greater and greater replication of memes. ". . . think of the memes as having created the Web to aid their own replication. . ." (pp., 216-217). Well then, memes seem in control of not only 'self', but also of the evolution of human societies and of the *nature* of the process of change (social evolution) itself.

However, when we consider these and other examples of memes in terms of their generating relatively persistent patterns of human social behaviour, the picture looks less impressive – and not a bit more convincing. Memes include: urban myths (drying the dog in the microwave), rules of political correctness, the habit of recycling bottles, religion, recipes, maps, written music, polyandry, singing, how to hold chopsticks, what to wear, how to say please, thank you, no thank you, altruism, the potlatch, alien abduction, computers and a theory/story of self. Not exactly the stuff which might be theorized into the causes of *significant* social or political behaviour, let alone of the 'forces of history'. Of those with some potential for significance (e.g., religion, altruism, the potlatch, and a theory of self, for example) little differentiation is made from the insignificant ones. More importantly, there is no indication of how they are derived from genes - most often it is denied that they are - or, on the other hand, how they might translate into, for example, relatively consistent human sexual fantasies, conflicts and competitions, or into marriage patterns, human groups, hierarchical relationships, politics or the state.

I doubt very much that this approach will ever be able to solve these problems. An *analogy* with evolutionary theory is no escape from teleology, or indeed from simple mysticism. An analogy is not cause, and, as I have argued, cause is essential to a realist approach to science, including a social science. Much of the work of Blackmore reads almost as if memes were some sort of devious, consciously conspiring, manipulating entities from a world not altogether comprehensible by normal humans. They remind one of the descriptions of witchcraft power as reported on by demonologists, anthropologists and historians alike (Cf., *Malleus*, 1486; Evans-Pritchard, 1937; Kluckhohn, 1944; Robbins, 1959; Midelfort, 1972; Kieckhefer, 1976; Anglo, 1977). There is a consensus in these works that witchcraft power was conceptualized as being a mystical power which infected the bodies of unsuspecting individuals. This power would then do great harm to individuals, society, environments and history , operating *through* the 'innocent' carriers of this power. Infected individuals had virtually no control over what happened. Blackmore does not usually suggest that memes cause harm - or good - (but gets very close when referring to the "viruses of the mind" idea – p., 22 and in her consideration of religion, for example). In fact, there is little suggestion that cause is involved (it is only mystically implied). But that is a large part of the problem; there is no sense of how memes can be conceptualized independently of biology or of how they might cause signifycant human socio/cultural behaviour.

So, we are back to the position of people such as Boas (1965, 1966), Kroeber (1917, 1952), White (1949) and Mead, (1942, 1943) whereby culture exists *sui generis*, almost God given, freely determining behaviour; the separation of body and mind remaining as great as ever. Only now it is even more vague and unstructured than when in the hands of its founders who did search for 'natural' patterns or configurations of culture, self-perpetuating cultural traits, typical sub-cultures, domain values or recurring structures, and so on, in an attempt to generate the basis for a theory of human society and history which did not picture human behaviour as being totally random. And also, often unlike the earlier approaches, most of what we know to be human is left out in memetics – at least in these latest developments. Lust, hate, love, envy, empathy, guilt, depression, fantasies, daydreaming, sibling rivalry, oedipal conflicts, deception, self-deception, avoiding, rejecting, bounding, bonding, status, hierarchy and politics, for example, tend to vanish from the analysis. Indeed, these kinds of activities and feelings have come to be seen as enemies of memes, destined to be overcome by memetic evolution. Blackmore, for example, makes it very clear that emotions, feelings, fears, cognition and perceptions are not memes (Cf., pp., 46-50) and are positively harmful in their contribution to a "false self"; guilt, shame, embarrassment, self-doubt, a fear of failure and the "selfish selfplex" (pp., 243-246).

We do not, in memes, have an advance on the earlier concept of the superorganic. We do not have a concept which links Darwinian processes in human evolution to 'shapes' of or motivators/propensities for significant human behaviour. Culture failed to provide this before, and it seems doomed to fail again as a master concept for explaining human societies and human history. Cultural studies today are much more concerned with studying youth behaviour, the modern media, advertising, sexual presentations of self, football culture, and so on. As worthy as these activities are, they are not exactly a search for a theory of mind, linkable to genes in one direction and human behaviour in the other. In any case, by the mid-Twentieth Century in the social sciences, culture as a master concept was being supplanted in most circles by a 'social determinism' in which both culture and social behaviour were said to be derived from *social processes*, if not social structures (these were discussed in terms of their inadequacies for linking genes and behaviour in the earlier parts of this chapter and in Chapter One). Unfortunately, this escape into teleology (as noted) took us even further from the theory of mind than culture. So, what about those scholars studying the working of the human mind itself, and resultant *individual* behaviour (largely psychologists)? What might they have discovered in the way of something like, perhaps, 'persistent patterns of mind'?; after all, they could not escape the problems of a human essence so easily as the above mentioned anthropologists, psychobiologists and sociologists had done.

Patterns Of Mind

But even for them human nature increasingly took on tame and rationalistic characteristics. George Kelly (1955), for example, developed a psychological theory in which instrumental rationality was combined with an almost mystical innate tendency for the continuous generation of mental "personal constructs". These were logical formulations of "constructive alternatives" which classified the world (often into binary opposites) giving individuals information which made it possible for them to cope, usually happily, with life. Human actions were simply based on humans "being alive" and each individual being a "scientist" by nature. There were no other motivations for behaviour; for Kelly the concept of motivation in terms of human behaviour was redundant.

Abraham Maslow (1968, 1987), on the other hand, suggested an innate, dynamic "hierarchy of needs", in order to add a positive dimension of 'satisfaction motivation' to Freud's repression motivated psychology. This hierarchy runs from basic biological drives to, at the highest level, a 'need' for "self-actualization". Unfortunately, for our purposes here at least, Maslow made little attempt to establish these needs non-teleologically, or to integrate them into a theory of social behaviour. In the end, self-actualization is tautological in that it states that 'individuals want to do what they want to do'; Maslow's needs were largely treated as teleological causes and also as a method for judging human behaviour in that, if individuals were not trying to achieve them, such individuals were said to be mentally unwell because of a lack of "achievement motivation".

Others looked to language (e.g. Chomsky, 1957, 1972), or to other forms of 'deep symbolic structures' (e.g., Levi-Strauss, 1963, 1966; see also, Habermas, 1971, 1984-1987), for forms of human 'pre-knowledge' 'concepts' which would allow their explanations to escape the teleological circles created by culturologists, social determinists and 'coping/needs' psychologists, without needing to give up rationality. The problem with these approaches, however, is that, so far at least, they tell us very little about actual human motivations. I doubt that they ever will. Deep language and/or symbolic structures are either sufficiently deterministic, in a robot-like sense, so that we do not need further explanations (that is, 'deep structures' are only carriers of pre-determined genetic messages), or they are only part of much larger emotional and cognitive patterns (see Chapters Five and Six).

The same criticism applies to C. Lumsden and E. O. Wilson's (1981, 1983) attempt to establish a relationship between genes, pre-knowledge and culture through their search for "epigenetic" rules. They might be right that human behaviour is directed, or at least strongly informed, by genetically pre-determined memory tendencies, just as Chomsky and Levi-Strauss might be right about the existence of deep structures, but the problem is just the same. Epigenetic rules are statements about pre-knowledge but they do not explain the proximate mechanisms and procedures which would be necessary to make their importance felt. Humans do not act as robot-followers of epigenetically determined memory information. Wilson and Lumsden come at the problem from the genetic end but, unless they wish to treat humans as mechanical followers of their memories, their reliance on rationality without emotion is no less than that of Chomsky and Levi-Strauss, who come at the problem from the cultural end.

What about language? Human symbolic and language abilities are very great indeed compared to other species, but it is difficult to perceive how humans might be motivated by the 'coldness', or innate neutrality, of language or symbolic patterns in and of themselves. Analyzing language in order to squeeze out meaning, for example, can be interesting but meanings do not cause behaviour, let alone represent scientific ‘truths’. Different individuals very often have different meanings for given words. Indeed, the same individual often uses somewhat different meanings at different times for many words in order to reflect their varying perceptions, fears and desires. That is why languages change. Quite apart from changes in pronunciation, meanings change in the minds of individuals, even without them knowing that it has happened. Agreeing meanings, however, does not result in knowledge, except possibly short term, for those engaged in specific interactions. Even then, individuals most often change the concepts they use to describe, or explain, their feelings ('personal truths') rather than change their feelings in order to agree with concepts.

Language is a tool of inter-personal interactions. It is very useful in communicating information, presenting self, deceiving others, advocating positions, defending self, being congenial and in seeking confirmation for rationalizations (Cf., Dunbar, 1996; Pinker and Bloom, 1992). Social life would be very different without it, human civilization might even have been impossible without it, but as a direct cause of behaviour because of its *intrinsic* nature, individual or historical, it has little power. We need to know why some concepts are emotionally charged and others are not, why a discussion about weather is usually boring and one about adultery is usually not. And why is the first not likely to lead to violence but the second one much more likely to? And why is reaching an agreement on the meaning of adultery not likely to alter the above possibility? Why is an attempt to discover human nature exciting and/or threatening, and an attempt to figure out the square root of 467 not? In other words, without *emotional meaning*, conceptual meaning, formative grammar and symbolic abilities do little more than classify the world.

In the final analysis - unless we wish to join straight genetic determinists (genes acting directly or acting robotic-like though deep structures or epigenetic rules, for example), on the one hand, or theologians with the soul/self being almost, if not totally, supernaturally generated and/or those social constructionists who declare human independence from biology as a matter of faith, on the other hand - we are left with the problem of discovering some "obligatory internal dynamics" (Panksepp, 1982) to explain why entropy has not won; that is, why the species *Homo sapiens* is still here.

The concept of pre-knowledge is important in this respect, indeed essential. But without some understanding of emotional, or some other similar motivators, pre-knowledge itself has little relevance. Human nature is not only about a species typical memory but also about species typical motivators/propensities. These must include the motivations for emotional and seemingly irrational behaviour as well as for reason and instrumental rationality; for hate as well as love; for envy as well as empathy; for fears and often seemingly insatiable desires and anxieties as well as for feelings of happiness and perceived tranquillity; these must encompass motivators for war as for peace; for sports as for dedication to 'work'; for selfishness as well as altruistic behaviour, and so on. In other words, the mechanisms we postulate must not only be capable of perception, calculation and memory but also, emotional-cognitive processing.

Fortunately, neural sciences and cognitive sciences have moved a long way since the earlier of the writers mentioned above were struggling with the problem, and it is possible, I will argue (see Chapters Four, Five and Six), to discover a physiologically and psychologically based "obligatory internal dynamic", or at least an outline of one (which incorporates emotionality as well as reason), which, in turn, can be linked to human social behaviour. Before attempting to do so, however, it is necessary to reconsider existing systems of classification because several of the categories generated by traditional systems have come to be more than heuristic devices in the minds of many biological and social scientists. They have come, in different ways, to represent independent, causative forces. As such they greatly obscure understanding of our biological base

THE PROBLEM OF CLASSIFICATION SIGNIFICANCE AND UNITY

Scientists (physical, biological and social) classify the material of their observations. "The basis of science is systematic classification" (Radcliffe-Brown, 1952, p., 7). Classifying (typology formulating) is a process of lumping specific actions, behaviours, beliefs and materials together, and then splitting them off from other lumps. The issue in classifying is: what should be lumped together and what should be separated? This is very important because a given classification system greatly affects the way the world looks in terms of what is considered to make up the independent 'essences' of the universe, what might be held to cause what and, of course, what are the significant dimensions of what is being studied (Hirst, 1976; Simpson, 1961; Rosenberg, 1981; Brown, 2000).

There is generally a psychological/socio/political element which invades the classifying process so that the universe often comes to be seen in terms of good or bad. For example, classifying can easily imply binary oppositions which suggest desirable and undesirable, progressive or regressive 'essences'. Categories (essences), then, when treated as causes of individual behaviour and of other essences, can be presented as elements for good or evil.

Social Scientists

Some social scientists (and philosophers and public reformers), for example, have created a category from concepts such as: reasoning, rationality, foresight, a sense of purpose and a drive for self-perfection, which they have generally considered to represent *humanness*. This 'lump' is opposed to *animalness* (instincts, tooth and claw, dumbness, beasts of burden) and, thereby, they proclaim a special, almost spiritual, uniqueness for humans. When analyzing human behaviour, social scientists often lump such things as reasoning, sympathy, empathy, trustworthiness, loyalty and self-control as being desirable aspects of 'the human potential', as opposed to a category which includes: passion, 'selfishness', un-trustworthiness, disloyalty and greed, all enemies of the human potential. A variation of this theme is to use some of these and additional behaviours - including an extensive capacity for learning and the development of philosophy and science - to create a category of rationality/reason and split it off from a category of irrationality/emotionality (including an unwillingness or inability to learn). Another variant has been to create two opposing lumps which divide a balanced/self-actualized/non-alienated ego from an alienated/anomic, unhealthy ego.

It is clear that something is being said here about human nature, human behaviour and human potential, and about what aspects of each are to be considered significant in explaining humans' place in nature. It is evident that a split between human and animal, between spiritual and biological, between good and bad behaviour and between good and bad types of humans is being portrayed. The above classifications also make it feasible, if not 'natural', to attribute 'needs' to the lumps (categories) based on the *definition of their natures*; and from this process, *causes* are implied because needs, by definition, seek to be fulfilled. For example, if as a category, rationally inspired sympathy, empathy, and so forth, is defined as humanness, humanness *needs* individuals to behave in the above manner. It is easy to believe, from this, that most humans will, in the end, be 'compelled' to do so, based, it is felt, on nature's 'hidden hand' of *naturalness*. Or, if a person does not behave as expected, it can be said that they suffer from an unbalanced, un-healthy ego. So, cause, prediction, good, bad, morality and immorality all grow out of systems of classification.

If we move from an analysis of humanness and individual behaviour to a consideration of society and culture, the same processes apply. In the early days of developing general theories of society, social scientists tended to lump together shared *symbolic* mental activities (cultures/ideologies), which they separated from social behaviour, and both of these from individual behaviour, sometimes known as 'The Individual'. In those days, a privileged one among these categories was often treated as causing the others. For example, culturologists used culture as supreme/ultimate cause of individual personalities and social interactions; early sociologists and British social anthropologists turned social behaviour into *social structure* as the cause of the others; and psychology generally used 'The Individual' as determining, at least individual responses to, cultural and social behaviour.

Again, the above approaches were communicating something about humankind and human history. As noted, culturologists were determined to show that humans were above biology (especially race) when it came to determinants of human behaviour. Humans were rational and were the products of symbolic *mental* creations, not base biology (Freeman, 1984; Harris, 1968). Part of the perceived task was to distance anthropology as far as possible from connections with evolutionary theory itself. For those looking at social behaviour, the view to be conveyed was that humans were basically social, cooperative, community orientated creatures. In the hands of some, society itself became God; society was the moral arbitrator (Durkheim, 1965, 1982; Radcliffe-Brown, 1952; for a discussion see Harris, 1968).

For those who wished to give a more material base to the notion of 'social systems' or 'social structures', the division of labour was given a sort of rational momentum of its own, one with determining power. Its behavioural results might have been exploitative in its formative stages, but its historical thrust was destined to bring eventual individual and social perfection (Marx, 1867; Radcliffe-Brown, 1952; Parsons, 1966; for a discussion see, Harris, 1968). Early psychology struggled to show that human behaviour was *learned* and that healthy behaviour and attitudes could be learned in place of unhealthy ones (Mead, G.,1962; Kelly, 1955; Cooley, 1902). Although they were less prone to seek out grand theory (and therefore not so quick to use 'The Individual' as a grand cause as culturologists or sociologists had been to use culture and society as their package of causes), psychologists were, nevertheless, keen to provide a scientific basis for the claim that rational, liberal, free individuals were natural. As part of this, they were often mistrustful of explanations which took cause too far away from the individual and gave it to abstract totalities.

As the social sciences developed during the early to mid 20th Century there was a tendency to split social and/or cultural activities into more specific categories: such as, for example, families, kinship patterns, institutions, classes, sub-systems, beliefs, values and ideologies. These categories were then portrayed as being in a feedback/causal relationship with each other in that each part was seen as being the cause of the whole, and the whole as the cause of the parts. The sub-categories were said to exist because together they fulfilled the 'needs' of a greater whole, 'society', and society existed to protect the needs of the parts. Societies, as a result, came to be seen as relatively balanced, synchronic wholes; not necessarily, it must be pointed out, working for the good of all. Marxists, for example, argued that the synchronic whole of capitalism functioned best for capitalists, subjecting others to exploitation; feminists argued that it worked best for males, and blacks that it worked best for whites.

But these observations were generally not taken as evidence against design in which functionally inter-related parts (as 'evidenced'/created in the systems of classification themselves) seemed proved by the existence of the synchronic whole. Rather, the view was taken that there must inevitably be a design which would satisfy all humans better than those now on offer. When the notion of inevitable progress was added, capitalism as a category, for example, must, Marxists argued, necessarily give way to socialism, as a category, and this eventually to communism as a category (admittedly at this point systems disappear). Here problems with identifying the functions of a social system in rational and progressive terms are deflected to another system of classification, a system of historical classification. Rationality and progress are not to be evidenced in the system of static social structures, but in history; from stage to stage, rationality and progress march forward. For their part, liberal and socialist feminists looked to social reform, as did most of the Civil Rights Movement in America, for a better, fairer social design *in the future*; what they did not do was challenge the notion of design.

There was, then, among social scientists of a general functionalist and reformist variety (largely sociologists), an accepted notion that functionally efficient social relations/structures represented a higher form of rationality. It was not uncommon that such systems were considered as having a moral force of their own. Durkheim's organic analogy and 'Society as God' idea may have faded, but 'scientific socialism' in the Soviet Union was a doctrine that 'the system' (based on Marxist/Leninist ideology) was unlikely to make mistakes; indeed, that it could not be wrong. This moral force was assumed to be greatly aided by reason and dedicated study, whereby individuals were expected to come to see and accept the value of the unity of purpose that the new 'society' was meant to represent.

But even where the supposed moral force of social structure, or the 'relations of production', were seen as being dysfunctional, the power of structure to cause individual behaviour remained very strong. For earlier Marxists, for example, since the functional unity of capitalist society was claimed not to be to the advantage of most of the population, individuals somehow naturally developed 'false consciousness', in which they believed that all was for the best in the best of possible worlds. So, they, too, were compelled by a 'hidden hand' of the structure, despite the fact that it caused them to behave in ways which were against their own best interest.

It is my argument that these forms of lumping and splitting are not useful because they make societies appear as part of a greater design rather than as a term to describe fallible humans living social and political life as best they can. There is often very little in any of the categories which might tell us how specific types of 'rationality', learning processes or cultures/social structures, for example, relate to the emotionally ambiguous, sometimes violent, often competitive, not uncommonly anxiety ridden, politically *charged* nature of human relations; characteristics so easily observed in all types of human societies. Often there is little indication in these approaches as to how the categories have come into existence, operate in terms of personal motivation and/or are changed. It is simply argued that these lumps *exist* as significant essences and that they have specified relationships to each other, and that they somehow cause each other because they *need* each other. But the generation of needs grows out of the tendency to attribute moral significance to certain 'essences' during the process of classifying and subsequent theorizing, not something that has a real existence in the world. This form of teleological explanation of contents of cultures, and functions of structures and structural parts, was the product of a desire to find out how human 'society' might be possible. It represented a desperate search for causes of 'structures' and, thus, hopefully, the conditions of stability, and eventually for the causes and conditions of progress.

Unfortunately, those who became critics of the rationalistic, teleological approaches of the classical social sciences (e.g., realists, phenomenologists, critical theorists, postmodernists), and questioned their systems of classification, did not discover or theorize answers to the fundamental questions asked by the classicalists concerning the nature of humans and of human social life; questions which we cannot ignore if we are to understand humans in any fundamental sense. Does this mean that it is not possible to break away from teleology in the social sciences? Or, are the claims of sociobiologists and psychobiologists that their new paradigm solves the problem, valid? Have they moved us forward?

Biologists

Despite Darwin's major theoretical breakthrough, biologists remained relatively non-theoretical for some time. Builders of systems of classification were kept busy refining geological categories depicting the Earth's history, and with locating newly discovered plants and animals into existing systems of classification; adding species here and there; and if they were *really* successful, a genus. Scientific arguments concerning whether various specimens of a particular genus should be considered three or five species, for example, abounded, with the vitriol associated with trying to discover God's will. Other types of classification had more to do with the development of academic disciplines in biology rather than theoretical debates. So anatomy, physiology, genetics and ecology each became a separate category. As areas of study, these were generally functionalist in the extreme. Anatomy was the study of animal and plant structures, physiology of how the chemistry of the organism functionally keeps it going, genetics was about maintaining existing structures through reproduction and in predicting what was nature and what was nurture on the basis of Mendelian mathematical calculations of inheritance, and ecology considered how it was that animals were so 'beautifully' adapted to their environments.

The development of bio-chemistry played havoc with this neat pattern because it suggested overlaps between physiology and anatomy. The study of embryology was always an uneasy fit because it looked to link physiology, anatomy and ecology into a *developmental* pattern; with the discovery of DNA, genes too were brought into this equation. So, by the 1940s – 1960s biology was facing a crisis of classification. What had previously seemed to be 'natural' fits with nature came into question. Sociobiology represents the first major attempt to re-classify biological material; it was an attempt to reclassify it in a *theoretical*, as opposed to simply a descriptive, way.

So, sociobiologists lumped together various electro-chemical processes of DNA replication - which demonstrates an incredible capacity for exact replication – with numerous ideas (especially mathematical constructs) from population genetics and ethology (which suggest that natural and sexual selection work at the level of individual organisms rather than at group or species level - 'the selfish gene'). Observations from ethology, and certain mathematical formulations from population genetics, further suggested to sociobiologists that Darwin's growing suspicion that sexual selection might be as, if not more, important than natural selection in the origin of species, was correct.

With this linkage, it appears that genes which provide for those reproductive behaviours which produce the most healthy and potentially reproductive offspring, will be most represented in a given population. There are, however, no specific genes which will necessarily do this in all situations at all times and places. What is consistent is that, within a pool of genes, those which provide more benefits for other individuals than for self do not long survive in a population. Thus the notion of relative 'inclusive fitness' in which there is no one characteristic or pattern of behaviour which definitely leads to the survival and reproductive success of a given individual, but rather that success is relative to the circumstances and elements of chance and luck which might activate a variety of reproductively advantageous genes. Thus, if the phylogenetic *location* of a species is fairly well known, its original reproductive behaviours can be postulated from traits shared with cousin species, and these then compared to the evolutionary outcome of those behaviours. So we can ascertain some of the specific behaviours which were selected for during that history. This suggests at least some of the types of behaviour in the current species which have genetic underpinnings (for example in humans, pre-knowledge mechanisms).

Thus, fitness theory - the new unity of 'selfish' genetic and individual behaviour, together with principles of sexual selection (centering around competitions for mates) - is a great step forward for neo-Darwinian theory in that it potentially displaces the teleological adaptionist approach, with its very unhelpful separations between anatomy, physiology, genetics and ecology. Fitness theory is quite an elegant formulation but, as we have seen, the problems begin when sociobiologists (and certain psychobiologists) try to develop this into theories of social behaviour, especially human behaviour. A mathematical/games theory-based category of human sociability, which includes, for example, kinship, nepotism, altruism and reciprocity, is generated in order to try to show how the selfish gene can result in relatively benign, rational human societies. The first unity (fitness theory) is treated as the cause of the second category, which in turn is treated as the cause of a third category, social structures/cultures.

The problem is that it is not shown how the first category links as a cause, or multiple causes, to the *contents* of the second (the sociobiologists' dimensions of sociability, each of which, as has been noted, are represented by largely descriptive rather than causative concepts); and the contents of the category of sociability do not easily link to complicated social behaviours, such as, for example, politics and *polis* formation (behaviours which might make up something approximating what we call 'society'). At the same time, very important emotionally based behaviours (such as seen in:

oedipus conflicts, sexual attractions/repulsions, buddyships, fears of strangers, vengeance seeking, homosexuality, being a fanatical sports fan, spirituality, nationalism fervently seeking 'success', personal politicking and war, for example) are lost out all along the way. This is because they do not easily lend themselves to the rationalistic - cost-benefit, mind as a computer or computational system, games theory - analysis these workers use.

Recent work in search of a Machiavellian intelligence (Cf., Whiten and Byrne, 1997; Byrne and Whiten, 1988; Dunbar, 1996) has considered some of the more conflict prone aspects of human social life - such as those found in deceit and counter-deceit and alliance formation, for example. This Machiavellian intelligence is formulated as a category in contrast (in part) to the 'kinship/reciprocity/altruism' category which has hitherto held sway. The emphasis in this type of work has been on the evolution of a rationally calculating mind, the better to win in social conflicts. This is helpful, but it still leaves out a lot of the emotional humanness which we have to explain. There is an overriding assumption that a greater and greater human intelligence was an advantage as humans lived in ever larger and increasingly complicated groups (which it was assumed was an advantage). It is far from clear, however, that larger and larger groups were an advantage for gathering and hunting, but, more importantly, one needs considerably more than a superior calculating mind to live in, and have reproductive success in, groups of *any* size; a mind too clever at calculating might, in fact, be a disadvantage while emotional comprehension could be a great advantage.

In Search Of A Bio-Social Classification

Neither sociobiologists nor psychobiologists nor social scientists appear willing, or able, to create a category for the purpose of analyzing all, or even the most significant, of the processes which exist between genes, on the one hand, and individually motivated social behaviour, on the other. The first two disciplines almost credit genes with selfish rationality capable of cost benefit analysis (sometimes operating as mind-as-computer), the third tends to jump directly to rationality by virtue of the fact that since the enlightenment humans have been defined as being rational creatures. To get beyond this, an essential question to start with would be, what mechanisms turn genetic propensities, discovered on the basis of criteria predictable from fitness theory, into species typical patterns of mostly learned behaviour? 'Rationality', socialization and functionalism from the social sciences, and the descriptive behavioural/social concepts and mathematics of population genetics and games theory from sociobiology, are of very little value in helping us answer this question without tautological and teleological propositions. Principles of sexual selection are more useful, but only as indicators of areas of behaviour where specific types of learning motivators/propensities might have been involved during human evolution.

To begin to answer the above question biologists might usefully seek to discover contingent connections between genes and bio-physical/chemical morphologies, especially neural structures and processes, includeing structures and processes which could contain pre-knowledge (brain 'forms' or 'knowledge units'). This would include an analysis of bio-electric processes, as they relate to the generating of knowledge units, dreams, images, memory potentials, daydreaming, thinking and behavioural motivations (such as emotions) in the human nervous system (Cf., Bloom and Lazerson, 1988; *Scientific American*, 1992; Greenfield, 1995).

Social scientists, especially psychologists, could lump emotions, reasoning and learning together (Cf., Buss, 1994; Pinker, 1998), and reconceptualize the whole package with a view to conceptually reconciling this lump with the above bio-chemical structures and processes. At the same time, social scientists could begin to consider contingent connections between this category and possible 'natural' (universal) patterns of behaviour ('natural kinds'). These connections would thus causally link processes of protein synthesis (genes and RNA) with such things as species typical memory units, desires, fears and patterns of self-identity formation (human nature), and these to potential natural patterns of behaviour.

Human nature would thus be the link (category of processes) between the work of biologists and social scientists. Human nature processes, and their expressions, it will be argued, are specific enough and sufficiently universal so that we can make generalized statements about them. Moreover, it should be possible to link them to behaviours without needing the teleology of the cost-benefit analysis which the sociobiologists' selfish gene usually entails. Social scientists, for their part, would need to accept the desirability of lumping emotions together with learning as species typical, universal, motivational processes, which would mean eliminating the distinction between animal and human, body and mind, emotionality and rationality. They would also have to consider splitting up their 'lump', which generally has societies as integrated, synchronic, functional entities, sometimes even treated as being both supra-personal and static in time and place, into smaller units ('patterns of behaviour').

This would be a return to the concerns of certain of the Enlightenment thinkers, especially the Scottish writers, to discover patterns of behaviour within what they called "civil society" (Seligman, 1992; Ferguson, 1966). The formulation of civil society at that time did not carry the same notion of design, structure and purpose that was later attributed to it as part of social systems or societies. Rather it was a concept in which individual actors, operating on the basis of a relatively fixed human nature, generated and operated through "fragile institutions" (Turner, 1993). Politics was often seen as a major means of conducting, organizing and, above all else perhaps, attempting to make safe human circumstances at all levels of human interactions. A concept of politics in civil society would also begin to make it possible to see human social life as being a rather random evolutionary process rather than as an almost supernatural movement to an ultimate design.

One of Darwin's (1859) major contributions was to present species not as fixed and specially created entities, but instead as populations of competitive, interbreeding individuals. When the time dimension was added, species became flows of reproductive competitions through time (Washburn, 1963a; Rosenberg, 1981; Brown, 2000). This made the elaboration of the processes of natural selection - contingent connections between reproductive patterns and environments - and the considerable understanding of organic evolution which followed, possible. It also condensed the processes of stability and change into one theoretical framework.

What we call human societies are, in a similar formulation, interactions of relatively constant species typical reproductive and survival emotions and cognitive processes in contexts of physical, social and historical environments. These emotions and cognitive processes are influenced by these environments but in turn exert a feedback effect for changing them. Living species and human societies can be compared to rivers. The apparent 'structure' of a river is a *temporary* result of the constant of gravitational effect on the relatively constant physics of hydrogen and oxygen molecules flowing through variable land conditions, the land conditions themselves being subject to change from the flow.

IN SEARCH OF HUMAN NATURE

If we are to get a scientific grip on human nature and proceed with a science of human behaviour, then, we will need to discover proximate causes (human nature motivators/propensities) of human behaviour and also reconsider our systems of classification (in the process re-examining the artificial separation between human 'rationality' and emotionality, and seriously questioning the existence of well integrated, synchronic, social systems). If we can do so, it will be possible to conceptualize human nature as one of Rosenberg's natural kinds. This will fill the gap between genes and natural kinds of 'patterns of behaviour' (tendencies to social and cultural universals, making up the ingredients of political society) and thus let us develop an empirically rooted science of human behaviour,- even given Rosenberg's stringent requirements.

In the remainder of this book I will be arguing that human nature is made up of an *integration* of cognitive abilities and an emotional charging capacity. Together these generate a number of basic motivating emotional-cognitive processes such as lust (and relatively more complex ones such as guilt), as well as more cognitively organized, highly charged 'species typical' desires and fears. In the process, it will be suggested, *charged* analytical categories are generated. Thus we are motivated to avoid some objects and behaviours and to seek others – those not charged are more or less ignored. Both the categorization and the charging tendencies are, to a significant degree, influenced, if not pre-determined, by the existence of templates (pre-knowledge) found in the human nervous system, especially the brain. Human nature also encompasses the storing of information (memory), which is retained in a charged manner so that 'emotional memory' acts as a feedback mechanism for future analytical categorizations and emotional charging activities. The accepting and rejecting of stimuli for storage is the human learning process, in which dreaming, playing, daydreaming, fantasizing, thinking and being entertained have a significant role.

Humans have also evolved a relatively good ability to categorize and imagine themselves in relationship to others (develop self-awareness) and, through the charging process, evaluate themselves (develop degrees of self-esteem), and evaluate others (attribute prestige and status). From this, it will be suggested, we get charged attractions (binding and bonding) and repulsions between self and others, and between groups of individuals. These, in turn, become a major basis of species typical social '*including*' and '*exclud-ing*' taboos and boundaries; they also become the basis of status hierarchies and human politics.

In this formulation we can imagine causation running from genes to social behaviour, but of course, in any specific situation causal forces move back and forth, sometimes rapidly, sometimes more slowly, jumping gaps, circling around, and so forth. Indeed, causation can operate right back to genes in so far as social behaviour acts as the selection force for the composition of gene pools. This formulation is meant to reinforce the point above that the human brain/mind is much more that a computational system: in so far as 'system' is the right word, it is an emotional system, an evaluating system, a classifying system, an anxiety prone system, and much more; not always systematic, not always operating in an individual's best interests – reproductive or otherwise

SUMMARY AND CONCLUSIONS

In this chapter I have suggested that although there are serious epistemological and ontological differences between biologists and social scientists, it is possible for them to work together and in the process to uncover a concept of basic human nature which motivates human learning, 'rationality' and social behaviour. Among biologists, sociobiologists have reopened the question of human nature and are well placed to postulate proximate causes of behaviour from a consideration of fitness theory and human evolution. At the same time, social scientists can seek universal patterns of behaviour in order to help in the search for proximate causes (human nature motivators). I doubt that the sociobiologists' notion of ultimate cause will be of much help in this matter. Fitness theory, as such, is a summary statement about a number of causes which have brought us to a certain point in phylogenetic history, not a cause of social behaviour.

I have argued that the major proximate causes of human behaviour are to be found at the level of human nature. Human nature represents the interface between relatively permanent DNA, with its specific messages for specific protein synthesis, and relatively variable behaviour. As such, human nature provides an area for study between the biological and social sciences which requires input from both areas. It also acts as a check on theoretical speculations in both areas. If social systems cannot be reduced to emotional and cognitive underpinnings (human nature), the theoretical frameworks explaining the systems are suspect. If genes cannot be postulated in terms of producing species typical emotional and cognitive proximate motivators the existence of such genes (e.g. genes for reciprocal altruism) is highly unlikely. Species typical pre-knowledge along with species typical emotional and cognitive processes make up human nature. The next step, then, is to begin our consideration of these processes in some detail.

CHAPTER FOUR
THE BIOLOGY OF EMOTIONAL-COGNITIVE PROCESSING

Human nature can usefully be conceptualized in terms of two basic dimensions: 1) emotional-cognitive processing and 2) advanced cognition. Since emotional-cognitive processing is usually held, phylogenetically and ontogenetically, to predate advanced cognition (Panksepp, 1990; Leventhal and Scherer, 1987), it will be considered first (in this and the next chapter). Following on from this, advanced cognition will be analyzed in Chapter Six.

Emotional-cognitive *processing* will include those bio-psychological processes involved in 'pre-knowledge', learning, memory storage and bio-electric (emotional) 'charging'. In other words, the notion of emotional-cognitive processing sets out to link neural physiology with what has traditionally been thought of as human emotionality and mood states. These latter, in turn, are investigated in terms of how they affect and motivate 'species typical' patterns of perception, learning, memory storage and the formation of emotionally charged desires and fears. As such, it is being argued that emotional-cognitive processes make up a large part of the link between genetic effects and behaviour.

The important point to remember is that the outcomes of emotional-cognitive processes will have to be explained as just that, *outcomes,* not as key or essential, elements of pre-existing or pre-designed moral categories, structural systems or inevitable historical progress. We are looking to discover independently operating emotional causative forces, evolutionarily derived, not teleological necessity or long term helmsmen of 'progress'. The argument, thus, will be that in terms of basic humanness it is not that we *need* something or someone but that we have evolved to *want* something or someone. We do not need to avoid certain things but we often do so because we fear them, sometimes obsessively. There is no (non-teleological) reason why we need a partner or to reproduce ourselves but we seek partners because we lust after and love certain individuals and/or we desire children; we do not need to worship sports and film stars, but get an enormous thrill from doing so. In terms of survival cost-benefit analysis, body decorations can be very difficult to explain but we are powerfully motivated to decorate ourselves, sometimes to the point of mutilation. Feelings of envy and jealousy hurt, sometimes terribly, yet we often cannot avoid them, indeed we often put ourselves in situations where they are likely.

Our conceptualization of human nature must take all of these activities and feelings, and many more like them, into account (just as it must consider how they might relate to reason, cognition and analytical abilities). Human emotionality, in other words, takes on a new significance; a significance as a prime dimension of the human motivational system.

EMOTIONALITY AND COGNITION: MEANINGS AND SIGNIFICANCE

However, just as Enlightenment-inspired social scientists were loath to accept a biological element in explanations of human behaviour, so too were they extremely hesitant to embrace emotions or emotionality as being anything other than a major hindrance to understanding and to 'right behaviour'. For numerous philosophers, and almost all social scientists, emotionality has been considered bad, if not evil, to be overcome with a strong will, rationality and clear thinking (Cf., Jarvie, 1984). Indeed, throughout human history emotions have often been given negative connotations, sometimes quite dramatically so.

Feelings such as lust, hatred, jealousy, envy and depressive feelings, for example, have been viewed as direct enemies of God, in the first instance, and then later of rationality. Such feelings are to be repressed, not embraced, let alone integrated into theories of human motivation. The usual human condemnations of lust, envy or jealousy are clear examples of the fear of human emotions. But even those emotions which at first sight seem desirable are often considered potentially dangerous when taken, or allowed to run, to excess. Consider the idea that too much love renders one a clinging person, that too much happiness is not good for the growth of the soul or for the development of balanced, responsible personalities; that a seemingly always happy person cannot be taken too seriously, and so on.

Furthermore, because emotions have often been considered biological in origin, and, as we have seen, because human biological and social natures have not uncommonly been considered to be at odds – biology representing an inability for personal control/bestiality with sociability representing control/humanity - it has been the tendency to consider it not just anti-rational but also *anti-social*, if not *anti-human*, to advocate an emotional basis for human behaviour. As we have also seen, in recent times this has led to a demonizing of some of the most serious attempts - as found in the work of Darwin, Freud and some psychobiologists, for example - to relate biology to emotions and emotions to behaviour.

To counter the notion of emotions as essential biological things with potentially evil/dangerous powers, some have sought to treat them as being 'socially constructed' (e.g., papers in Harre, 1988). This renders emotions subject to being 'socially un-constructed' and 're-constructed'. It also renders them subordinate to rationality and also to moralities, ethics, psychotherapy and 'proper' cognitive functioning. However, if emotions were simply the same as socially or culturally constructed attitudes and beliefs, we would not have felt a need for a separate concept of *emotions,* moreover one that conveys the *power* we feel they have. Nor would we have had a long history of fearing the inherent dangerousness, indeed, potential evil of emotions.

Cognition, on the other hand, has been seen as a capacity for advanced reasoning, d*esigned* to understand a rational universe. When behaviourism and experimental psychology (experimenting with rats, pigeons and sometimes children) did not seem to uncover the deterministic laws of its early promise, or to finally replace Freudian psychology, *cognitive* psychology came into its own. This trend was supported by humanists generally, but especially humanist oriented psychologists, who were beginning to worry about humans having been left out of psychology. More recently, as we have seen, certain psychobiologists have latched onto cognitive sciences to create a picture of mind-as-computer as part of a return to the supposed rationality of adaptationism in evolutionary thinking.

The approach of cognitive psychology so far has been to uncover patterns of human problem solving, based on rationality, first and foremost, with emotions thought to play virtually no part in explaining behaviour. "Cognitive psychology attempts to sharpen our understanding of the way mental activities work" (Benjafield, 1992, p., 8); he goes on to point out that in this discipline, emotions are often not even included in texts (Benjafield, 1992, p., 9), or at best have only been put in as a sort of add-on chapter (Cf., Benjafield, 1992; Eysenck and Keane, 1990; Forgas, 1981). The emphasis has been on problem solving through such things as information processing, reasoning, judgment, choice, language, intelligence, creativity and concept formation (see note three, p., 234). "The affective, motivational character of dealing with the social world is currently largely bypassed in research on social cognition" (Forgas, 1981a, p., viii).

Yet, it is exactly at the point of interaction among emotional and cognitive processes that biology and behaviour are linked in the human species; it is the *locus in quo* in which genes have their effect on consciousness, unconsciousness, behaviour, attitudes and social patterns. Emotions, in this sense, represent the cauldron of humanness, the force and momentary direction - motivation and guidance - behind perception, problem solving, creatvity and information processing. In fact, there has been a growing recognition among cognitive psychologists of the importance of emotions in understanding consciousness, including its motivational aspects. Increasingly it is being argued that in the actual study of human consciousness it is virtually impossible to separate emotions and cognition. This view might be summed up by Leventhal and Scherer (1987) who have noted that it is

> ". . . extremely rare to find emotional reactions totally separate from perceptual or cognitive reactions in the human animal. Indeed, it may be difficult if not impossible for a human being to experience a truly free-

floating emotion except in those rare situations that elicit only sensory motor processes and "emotion like" reflexes. . . Hence, "emotion" and "cognition", as labelled . . . are always intertwined in emotional behaviour and emotional experience" (p., 23).

Moreover, in recent years there has been some recognition that

". . . the absence of a motivational theory underlying current cognition paradigms is a serious problem. It appears as if we believed that the single overriding passion of people in everyday life is to achieve rational understanding " (Forgas, 1981b, p., 264).

Therefore, a desire to integrate perspectives concerned with emotions and cognition has become increasingly manifest (Cf., Crook, 1980; Buck, 1984; Buss, 1992, 1994; Nesse and Lloyd, 1992; Pinker, 1998). I strongly support this approach and hope that the attempt in this work to link biology and mind, genes and behaviour, will contribute to that effort.

Emotions, when linked to cognition, clearly are complex interactions of biology, psychology, behaviour and thought (see note three p., 234). In outcome they activate a considerable amount of physical change and energy expenditure directed against an individual's physical, social and mental environments; through them very powerful, long lasting memories are formed. As such they can be considered significant motivators of human behaviour. Jealousy, for example, might not be a pleasant experience for those who feel it, or for those who are subject to the behaviour caused by it, but it is, nevertheless, a basic force, which appears universal among humans (Wilson and Daly, 1992), and which motivates a significant amount of human behaviour and underpins a number of extremely important social taboos and boundaries (Van Sommers, 1988). The same claims for motivational power and universality can be made for a number of other emotional states, both those which are often condemned and those which are not. To see emotional-cognitive power as anything less is not only to deny (ignore) a large part of the humanity which has been with us, for better or worse, through the ages, but also to miss an opportunity to scientifically further our understanding of the motivators which lie behind human behaviour.

Although somewhat inelegant, the term *'emotional-cognitive processing/processes'* will be used to depict the basic workings of human nature. This is because we are talking about complex *processes* not essential *things*. Emotional-cognitive processes represent the evolution of an interaction between what have traditionally been classified separately as specific named emotions or as emotionality on the one hand, and cognition and/or learning and rationality on the other. It is important to note that we are not simply linking named emotions with named cognitive activities. We are looking to a major re-conceptualization of this whole area in which emotionality as processes looms large and emotions as named phenomena disappear. As such existing concepts which have the capacity to suggest fusion are highlighted and/or new ones emerge.

Human emotional-cognitive processing as a major dimension of human nature will be divided into three main sub-dimensions for analysis: 1) the biological nature and mechanisms of pre-knowledge, 2) the bio-electrics of acquiring and retaining 'charged' motivational knowledge, and 3) the psychology of emotional-cognitive processing. In the rest of this chapter I will consider the first two of these, the psychology of emotional-cognition will be considered in the next chapter.

THE BIOLOGY OF EMOTIONAL-COGNITIVE PROCESSING

Pre-Knowledge

Knowledge, or information, will be taken to mean stored 'guidelines' and motivators for potential future energy expenditure. Pre-knowledge can be considered genetically created 'messengers' lurking to inform the development of knowledge as embryological and post natal neural developmental processes are affected by experiences. The concept of *pre-knowledge*, in other words, represents the relatively fixed genetically induced neurological basis upon which the more variable emotional charging and cognitive ordering of acquired knowledge operates. Pre-knowledge, in this sense, helps protect the basic integrity of the messages of human genes sufficiently so that there remains the biological continuity which is represented as the species Homo sapiens.

In the first instance specific human biological capacities for perception act as gatekeepers to human knowledge. We observe a restricted number of phenomena because our sense receptors only accept compatible inputs. At the same time, from this restricted input, our sensory system (in conjunction with other forms of pre-knowledge) makes it almost compelling for us to experience certain things because it finds some stimuli extremely congenial. It must be noted, however, that it is not necessarily the case that we are consciously aware of all (or even most) of what we do perceive (Cf., Benjafield, 1992; Eysenck and Keane, 1990), or of the impact of what we do perceive. The point is that whether we are conscious of it or not, during evolution our receptor apparatus evolved to be able to generally sense signifycant reproductive and survival knowledge/information for evolving humans, rather than take in any and all - some potentially harmful - information (Gibson, 1979).

As important as selective perception is for the relative success of a species, however, it is the mechanisms which provide *guidelines* for the acquiring and *charging* of knowledge that make up the most significant basis of human pre-knowledge. What we are looking for is a genetically derived neurological basis for *specific* knowledge 'units'/bundles which are capable of motivating and directing the learning of what we can, as a result, consider to be species typical behaviour. It must be noted that while such knowledge units might be analytically distinguished from developing (the processes of learning) and already developed (memory), knowledge units, the distinction in terms of their interaction and effects is probably far from clear..

Although the bio-electric bases of pre-knowledge units are still largely unknown, it is hoped that with the rapid rate of progress in the neural sciences characteristic of the past several years this situation will soon be remedied. What does seem probable at this stage, however, is that pre (and developed) knowledge units are based on bio-electric action potentials (Cf., Fischbach, 1992; Bloom and Lazerson, 1988; Greenfield, 1995) loosely organized and stored by the neural processes which take place within the human brain (Cf., Kandel and Hawkins, 1992; Ornstein, 1991; see below).

Among the candidates for containing organizing potentials (pre-knowledge units) are the vast numbers of proteins, including neural transmitters, synthesized during general neural activity. Just as DNA directs RNA synthesis in such a way that its 'messages' guide the building of the complete morphology and physiology of an organic whole (with pre-programmed, often considerably delayed, results) so too could neurally generated proteins, or clusters of proteins, contain potentials which regulate neural transmissions so that they behave in a more or less pre-determined way. (The on-going synthesis of proteins could, of course, also be affected by internal - memory - and external stimuli. As such they could be a basis for developing and storing knowledge units, that is, for learning and memory. This would especially be the case if particular protein synthesis affected future synaptic structuring - Bloom and Lazerson, 1988.) The generation of both pre-knowledge and developing knowledge units through protein synthesis and clustering of proteins might be a function of specific neural transmitters or be a more general characteristic of most neurally generated protein.

Another candidate for being the store of pre-knowledge units, and certainly for becoming developed knowledge units, is the formation of neural circuits. Although the pattern of most circuits is set down by our genes, dendrite and axon growth and branching, and specific synaptic connections, only take shape after birth. (Visual experiences are very important in this process - Shatz, 1992; Zeki, 1992; Ornstein, 1991; Bloom and Lazerson, 1988). Especially important here, we can speculate, are 'single source/divergent circuits'. These are characterized by their ability to interconnect with a variety of systems and thus to function in integrating the many activities of the nervous system (Bloom and Lazerson, 1988; Greenfield, 1995).

Single source/divergent circuits involve specific clusters of neuron interconnections. This provides the possibility for storing complicated information (images, concepts - Damasio and Damasio, 1992) while, at the same time, maintaining the capability for affecting a multitude of other clusters through diverging neurons. Moreover, the transmitters associated with these produce "condition" transmitter actions (Bloom and Lazerson, 1988). This means that the results of the actions depend on the conditions under which the actions take place (see also Greenfield, 1995) and are thus amenable to environmental influences as potential triggers. And furthermore, single source/divergent circuits, although only a fraction of the total number of brain circuits, are concentrated in the limbic system and typically generate monoamine transmitters, that is, the area and transmitters most implicated in the human moods which, it will be argued, 'charge' human knowledge units (see below).

Another possible candidate for being the storehouse (or raw material) of pre-knowledge - and for developing knowledge - is the multitude of glial cells which fill the spaces between neural tissue. Glia, unlike neural cells, have the capacity to subdivide to form new cells, perhaps thus maintaining a consistent amount of an individual type of protein-stored information (potentials) through time. Some glial cells are specialized for rapid conduction of electrical impulses, making certain kinds of very rapid action possible. This also, perhaps, includes a rapid retrieval of complicated, relatively fixed (that is, rather specifically neurally configured) information. Rapid retrieval of knowledge also suggests the possibility of unconscious motivations in so far as the motivational force would most likely by-pass those parts of the brain where conscious analysis might interfere. Glial cells also 'clean up' excess transmitters and ions in the spaces between neurons, preventing electrical interference with firing rates and thus helping to regulate synaptic functions, and as a likely result, moods. More importantly, however, glial tissue is necessary for directing many neurons to their ultimate destinations and in the killing off of neurons 'considered' surplus to requirements. Although neural migration and growth are affected by environments, the pattern in which glia directs them seems largely set down by RNA messages from our genes (Bloom and Lazerson, 1988).

A further possibility for the brain areas which may contain, or at least process, pre-knowledge units (and developing knowledge units) are the columns which develop in the cerebral cortex (Cf., Mountcastle, 1975, 1978; Greenfield, 1995). Columns are made up of "minicolumns" which in turn are made up of one hundred or so vertically inter-related neurons that span the layers of the cerebral cortex. Connections develop between minicolumns and thus between columns, as well as between these and many other parts of the brain. Columns, and groups of columns, specialize in a variety of activities - such as in fixing gaze and visual attention, for example

- thus providing a basis for directing behavioural and/or learning potentials though intensified concentration.

They process information not only from external sources (environmental) but also from internal sources (memories, emotions, cognitive skills, pre-knowledge units). The latter is referred to by Mountcastle as "reentrant information" and when juxtaposed with external information may provide the basis for conscious awareness. That is, in this capacity, cerebral columns may well provide a basis for separating self (internally oriented columnar organization and/ or activities) and others (externally orientated columnar organization and activities). From this there is a basis for the constant evaluation of environments in terms of 'self'. And, given that columns tend to operate hierarchically, there is a basis for the establishment of 'command' levels from which an "I" can not only evaluate but also direct reactions to environments (Mountcastle, 1975, 1978).

Whatever the case, be it one or some combination of these, or some mechanism not yet discovered, there are ample genetically derived brain mechanisms which could easily provide for pre-knowledge units (and the developing of knowledge units as well). Brain composition in this regard shares similarities with elaborate arrangements of computer chips; knowledge units and potentials for developing knowledge units being in some ways organic equivalents of both computer chips and software. Both are capable of storing information for directing the future use of energy. That is, both are capable of carrying and directing electric impulses in search of particular knowledge unit compatibility and/or congenial receptors in order to activate stored energy and direct it into specific motor activities.

The human brain, however, is not a computer and its mechanisms are much more than an arrangement of chips or software (Greenfield, 1995). The human brain not only processes information but also evaluates and re-evaluates it, classifies and re-classifies it, manipulates it and, very significantly, emotionally 'charges' much of it. And it is these characteristics, rather than mechanical (electronic) information gathering and processing as in a computer, which result in the motivating of human behaviour and in the storing of guidelines for future behaviour. This brings us to emotional-cognitive processing 'proper' - to the bio-electric processes in which pre-knowledge influenced learning and the 'charging' of knowledge units (and eventually the creation of species typical desires and fears) takes place.

Acquiring Knowledge

Learning and bio-electrical charging also take place largely in the brain. The one hundred billion, or so, neurons which make up the brain and the rest of the nervous system have two major potentials: to be either electrically excited or inhibited. Excited neurons send messages (neural transmitters) to each other; these in turn may further excite or they may inhibit electronic potentials (firing rates) in other neurons. Some neurons have as their major function the modulating and inhibition of excitation in others. Hormones generated in the endocrine system also affect the receptivity and excitability (firing rates) of particular neurons.

Of key importance are the activities of neural transmitters because they carry the information which alters moods and directs reactions to stimuli. They include, for example, amino acid transmitters for the organization and activation of physical activities, monoamine transmitters (e.g., norepinephrine, acetylcholine, serotonin and dopamine) for the development of pleasure, mental concentration, sensitivity and tranquillity states, and peptides (such as enkaphalins and endorphins - endogenous morphines) for the easing of tension and pain. A particular 'mix' of firing rates of neural transmitters and endocrine hormones affect our physical behaviour, abilities to concentrate, states of arousal, sleep patterns, social behaviour, and, very significantly, what we learn and how we learn it.

Neuron firing rates can be said to operate as a dynamic balance, manifest as fluctuating *mood states*. Moods, for example, can swing from highs to lows, sometimes from mania to deep depression (Gershon and Rieder, 1992; Gilbert, 1989). But as a rule genetically programmed and/or externally influenced regulators (e.g. amino acid transmitters) modulate neuron firing rates and thus maintain balances (Mahendra, 1987).

Learning

During these activities considerable protein is synthesized within the neurons and, thus, future 'action potentials' established. Sometimes relatively permanent connections are created between neurons so that "gestalts" (Cf., Greenfield, 1995) or 'forms' or 'structures' of them are established (circuits created). In all these cases the potential for the longer term storage of information exists in a context in which stimuli (both externally and inter-nally generated) are setting in motion the excitation or inhibition among neurons which, it can be suggested, give knowledge units their motivational ('mood') charge.

Memory units, as with pre-knowledge units, most likely consist of newly synthesized neural protein, altered synaptic interconnections, new synaptic interconnections or cerebral columns (or some combination of these). Perceptions are processed and some of the results stored briefly in the hippocampus and amygdala. Rather fewer of the results are stored for a longer period of time in the neocortex. Bio-electric charging (see below) takes place in the hippocampus and more analytical activities (see Chapter Six) take place in the cortex, with feedback loops between them. Conversion of some of these into long term memory most often happens in the neocortex. A major candidate for providing the transmitter *charging*, which helps determine what will be retained and what will be discarded, are adrenal norepinephrines (Bloom and Lazerson, 1988), but other monoamines are also very important (Cf., Greenfield, 1995).

Knowledge units, then, are activated, shaped, filled out, created and charged in response to stimuli. Stimuli are more or less easy to receive and to process into more durable units. This depends on the nature of our sensory equipment and the pre-existence or absence of particular 'templates' (pre-knowledge potential) for knowledge units. Retention of information also depends on the amount and type of neural inhibition or excitation generated during its perception - that is, on the amount and kind of mood excitement and heightened sensitivity involved during perception (some combinations of excitement and heightened sensitivity resulting in intense involuntary concentration and other combinations in an inability for concentration, for example).

Some stimuli are so difficult to receive, process or retain that a considerable amount of induced reward is necessary in order to create knowledge units. This might be done through a type of teaching which simultaneously generates the same kinds of excitement and heightened sensitivity as more naturally develops during the learning of other, more congenial, things. In some cases, considerable voluntary (if not forced) concentration may be necessary in order to develop knowledge units. It is possible, however, that voluntary concentration can, after a time, activate the same reward mechanisms which are associated with involuntary concentration (Crook, 1980; Buck, 1984). The amount and consistency of stimuli are important, as is, of course, what has already been retained, in determining the ease or difficulty of maintaining and developing knowledge units.

Many of our species typical pre-knowledge units and learning potentials are probably activated and/or more or less developed during the first fifteen or so years of life. This is because it is during this period that neurons develop their dendrite branching, axons grow and basic synaptic interconnections are laid down. Early learning is especially the case with procedural knowledge (e.g., how to walk, how to eat) which does not use the cortex, but rather is largely controlled and stored in the cerebellum. But it is also likely to be the case with such things as, for example, language. We very easily learn to speak a first language but after the age of about 13-15 years we find it extremely difficult to learn a second one. It is as if after a certain number of language related neural interconnections have been made, the majority of future 'language stimuli' travel down those channels so easily that they do not readily fill out or create other potential language units. Indeed, these units may be able to generate a blockage on the development of random language units. This does not mean that new languages cannot be acquired (clearly evidence suggests that they can) but only that gaining knowledge after a certain age can be very hard work indeed or is more likely to be filling in the details rather than setting the format of knowledge.

It does seem that knowledge after a certain age is largely technical and procedural knowledge (the rules of football, how to use a computer, how to live in polite society, what *specifically* gains attention and the laws of the land, for example) and comes from on-going experiences. Original knowledge (running, kicking a stone - ball - lusting, loving, feeling envious and jealous, desiring attention, learning a first language, sociability) on the other hand, comes very early, and is greatly influenced by pre-knowledge. If the emotional-cognitive processes involved in the development of strongly felt attitudes, presentations of self and spiritual beliefs, for example, are the same as in language acquisition, knowledge units related to them might also be set down at a very early stage and tend to block the development of radically new ones.

It can be further suggested that much of species typical knowledge generating (learning) is largely an unconscious, and/or semi-conscious process. From this notion we can speculate that it is during such things as dreaming, daydreaming, fantasizing and playing that the most significant of our species typical knowledge units are formed. It is not so much that these activities are unconsciously or involuntarily entered into that is important - sometimes they are, sometimes not - but that during them a great deal of unconscious and involuntary learning takes place. These activities, thus, are not just frivolous activities to be ignored by scientists but are key processes in learning.

Dreaming may include an unconscious 'coming out', as dream images, of certain pre-human knowledge units (Sagan, 1977; Freud, 1976). More significant is the evidence that during dreaming there is a sorting of recently received stimuli and temporarily created knowledge units (Cohen, 1979; Hobson, 1990; Sagan, 1977; Bloom and Lazerson, 1988; Miller, 1987). During the high degree of protein synthesis and general neural activity which takes place during dreaming, argues Cohen (1979), some temporary knowledge units are stored as long term memory. REM (rapid-eye-movement) sleep - dream sleep - ". . . serves to reorganize our higher nervous centres according to some genotypic blueprint" (Jouvet, 1973, p., 31. see also, 1974). During REM sleep the same brain waves, amino acids and neural transmitters which are associated with heightened sensitivity in the wake state, are also present. These, according to Cohen (1979) are largely serotonin and norepinephrine. Recent evidence, however, suggests that acetylcholine could be much more important than serotonin, which may, in fact, shut down dream sleep (Greenfield, 1995). Whatever the specifics, the brain is very active during dreaming while physical systems are shut down. Thus all energy can be diverted to the sorting and storing of new information rather than to motor activities.

The right hemisphere and limbic system of the brain are deeply involved, with a loss of control by the more analytical left hemisphere, during dreaming. It is in the former locations that much of the image (knowledge unit) sorting and charging necessary for the development of long term memory take place. Dream sleep is, in fact, in sharp contrast to delta, or deep, sleep. Deep sleep is related to left hemisphere control and seems to be a complete shut down (extreme neural inhibition of both emotional-cognitive and physical systems) for the purposes of body restoration (Cohen, 1979); although some learning may in fact take place even then (Greenfield, 1995). Cohen suggests that, in the wake state, visual sensory inputs travel via the visual cortex area to the lateral geniculate bodies and then to the pons (which has a very high concentration of single/source divergent circuit neurons and monoamine transmitters) where they are directed to the cerebral cortex. From there they move through the thalamus to the motor cortex where they activate physical action. However, if derived from recently acquired, temporary, internal knowledge units (short term memory), and blocked from the motor cortex during REM (or other) sleep, at least some such inputs move from the pons to the occipital cortex to be stored as long term memory.

The importance of dreaming for learning is highlighted by the fact that the same physical and mental processes are involved as in conscious thinking and the triggering of motor activities, except that the motor activities are blocked (Freud, 1976 - Chap. VII; Cohen, 1979; Hearne, 1986; Green and McCreery, 1994). During dreaming, however, it can be argued that in the place of motor activities we get the powerful bio-electrical *charging* essential for learning. Moreover, dreaming depends very heavily on visual forms (images). And the importance (compared to most other species of mammal, and indeed even to other primate species) of visual perception in the development of human knowledge and, indeed, consciousness, probably cannot be over-estimated (Crook, 1980; Glass and Holyoak, 1986; Shatz, 1992).

History and our own life experiences, for example, clearly teach us that visual imagery has been very much involved in self-awareness, sexual behaviour and interpersonal and hierarchical/political (social) relations. These are all areas, it can be argued, that were under especially strong selection pressure during human evolution (see note four p., 234). Indeed, during the evolution of the primate brain the visual cortex took on many of the functions previously performed by other sensory and brain systems and, in humans especially, became the basis for the development of a large part of the neocortex. Since the neocortex is a key morphological area in terms of both the storing of information and advanced cognition it is possible to theorize that visual evolution represented a major part of human evolution.

For example, most humans have very well developed visual intelligences and memories. Almost all humans find it very difficult to remember names but rarely forget faces, can visually recognize friends instantly even at a distance and can very often determine moods of others by simply looking at their facial expressions and mannerisms (Ornstein, 1991; Ekman and Oster, 1979). It is quite common for small parts of objects or scenes to evoke visions, indeed understanding, of wholes (Watzlawick, 1978; see also Pinker, 1998). It is perhaps no coincidence that human eyes are often considered to be mirrors to human souls. Sexual partners, long term mates, friends, heroes and political protectors are chosen very largely on the basis of their appearance and behaviour as ascertained by our visual sense - directly or through dreams and daydreams. Very often too, we attribute value to objects and individuals on the basis of their physical appearance and/ or our visual imaginations of them.

From an evolutionary point of view we can speculate that visual intelligence would have been selected for as a means of recognizing the intentions, moods and deceitfulness of social others. From this point of view, then, it seems extremely likely that visual processes, such as potentials for dreaming, daydreaming and visual thinking, would be closely tied to evolving pre-knowledge and other potential knowledge units. But, as a by-product of the evolution of visual mechanisms, the capacity for increasing cortical interconnections able to store new human knowledge also evolved. Certainly, in humans, visual experiences seem to be very significant in the development of a number of congnitively important neural interconnections (Bloom and Lazerson, 1988; Glass and Holyoak, 1986; Shatz, 1992; Zeki, 1992). In fact, it is thought that much of this processing in humans may well take place as part of an interaction between the visual cortex and the rest of the brain (Bloom and Lazerson, 1988; Glass and Holyoak, 1986).

Nevertheless, the products of dreaming are by definition largely unconscious activities, and it is very difficult for us to know exactly what role they play in learning (Cf., Miller, 1987). Although there is an abundance of protein synthesis (moderators for storing potential) and of monoamine transmitters (charging potential) there is a lack of coherence in dreams. Serotonin injections can induce REM sleep but their receptor sites are also the ones which attract the hallucinogenic drug LSD. Imbalances of monoamine transmitters are implicated in schizophrenic and depressive conditions (Mahendra, 1987; Gilbert, 1984; Gershon and Rieder, 1992). Yet, on balance, it is hard to believe that there is no connection between dreaming, or at least visual imagining, and learning. The very fact that the above biochemical effects are related to the way we perceive and experience the word suggests a connection between them and behaviour. It also suggests that we are looking for the evolution of some kind of bio-electric 'balance' among these processes as the basis of what we consider to be humanness.

If we follow this line of reasoning it will take us somewhere like this. The next phase from 'learning from dreaming' (largely unconscious/pre self-awareness) in a human evolutionary scenario, could have been the 'remembering of', and 'adding to', dreams when awake. At this stage we are talking about daydreaming; daydreaming may have been the earliest manifestation of conscious human self-awareness. That is, daydreaming implies an ability to consciously experience visual images which are perceived to be separate from self's current existence. This separates self from other existences. In so far as these other existences contain other individuals, self is separated from others. When an individual can imagine other existences for self, they have the ability to locate self in the past and in the future.

An ability to arrange and re-arrange daydreams could be said to represent *thought* processes. These would be perceived to be separate from immediate sensory experience. Daydreaming and thinking, as an evolutionary development of the dream pattern, could involve a semi-conscious (if not conscious) diversion of proportions of sensory input from the pons through the left brain hemisphere for ordering (in relation to self) before sending them on to the occipital cortex for storage.

Of special interest for current research are those individuals who can control their dreams while asleep, so called "lucid" dreamers (Green and McCreery, 1994; Hearne, 1986; Garfield, 1976). In lucid dreaming semi-conscious control of knowledge units, largely in the form of visual imagery, is generated both before and during dream sleep resulting in a great deal of tranquillity, if not heightened sensitivity and desire fulfilment. Indeed, lucid dreaming is characterized by a pleasant, happy, interesting and often "exhilarating sense of adventure" (Green and McCreery, 1994, p., 47). This suggests that the evolution of the human ability to manipulate and even control visual images reached a point, at least in some individuals, at which it can be, to some extent, imposed back onto the more unconscious activity of dreaming. Whatever the case, the emotional intensity, sense of realism and long term retention experienced by lucid dreamers provides further evidence of a link-up between dreaming, emotional-cognitive charging, self-awareness, learning and memory.

Most people are not lucid dreamers, although in many cultures attempts are made to control the contents of dreams (Garfield, 1976). In all cultures there are conscious efforts made to return to the dream state. These include a number of institutionalized activities which usually involve an element of some form of intense daydreaming; as happens, for example, during fasting, prayer, rituals, yoga, hypnosis, meditation, spirit possession, listening to music, drug taking and participating in and watching some forms of entertainment. Even if everyone does not work as hard as this at daydreaming, everyone does daydream. Although operating in different conditions, lucid dreaming and daydreaming would both involve a degree of left hemispheric sequence ordering of stimuli at roughly the same time as a limbic and right hemispheric bio-electric charging was taking place. In daydreaming this would happen during the wake state, but with motor activity held to a mini-mum so that there was little diversion.

As a hypothesis, it is possible to suggest that we consider dreaming, hallucinogenic experiences and perhaps 'delusional depression' as being phy-logenetic remnants of an earlier stage in the evolution of enhanced visual perception, visual imagery, daydreaming and, eventually, self-awareness and thinking (see also, Freud, 1976, Chap. VII). The result has been a consider-able capacity for both the long-term storage of complicated information and for more controlled analytical thought. It can be further suggested that as this system evolved, daydreaming and thinking, and an integration of day-dreaming with the phylogenetically long standing role of play in learning, became the major methods of human species typical learning. This would be instead, perhaps, of a greater role for dreaming and/or simply play as such in the early stages of this evolutionary scenario.

Indeed some types of advanced knowledge are probably still more easily gained by modern humans during physical activity (play) than through the more mental processes of dreaming and daydreaming. Csikszentmihalyi (1975) defines 'deep play' as those activities which generate intense con-centration; many physical play activities do just that. Moreover, it is likely that physical play involves many of the same processes of bio-electric charg-ing as daydreaming because heightened sensitivity and enjoyable excitement can be very high during both. Play is a natural human activity and it is very unlikely that its phylogenetic role as a stimulator for learning would be lost.

During play both external stimuli and stimuli from internal knowledge units automatically activate motor activities. This often happens in a highly random manner (observe very young children at play). But it may not be as random as it appears. Clearly, only certain stimuli (internal or external) acti-vate play. This is due to the selective nature of our sensory systems, pre-existing knowledge units and the build up of specific emotional-cognitive motivators. Human play becomes less apparently random and increasingly ordered and stereotyped as more and more pre-knowledge influenced 'game plans' are developed. During play visual images and other knowledge units are charged and stored. At the same time, certain physical activities can be called upon to aid in daydreaming and thinking. Thus we often drum our fin-gers, tap a pencil against our hand, doodle, rub our hands, make all sort of facial expressions, knit, pace the floor, go for walks and jog, for example, to aid our daydreaming and thinking.

In a sense, then, while it is most likely that during primate evolution play was a precursor to daydreaming and thinking, it can be argued that it is only when the dimension of daydreaming was added that species specific 'human play' came into existence. For example, physical play might be the key to becoming a successful predator, or in humans for becoming a footballer and for developing competitive feelings. But an integration of play and daydreaming is necessary for becoming a clever footballer able to win by the rules of football, and for becoming a football coach to these same ends. Daydreaming on its own is most important in motivating a coach to develop a season-long game plan.

Species specific interactions of play and daydreaming among humans can be quite complex. While play can encourage daydreaming it can also restrict it. Dancing can generate romantic daydreaming but failed courtship play can slow it down. Playing ball in a field with mates can encourage daydreaming, playing against professionals can end such daydreams. Writing a piece of instrumental music is a clear example of the intertwining of daydreaming and play; interestingly here the daydreaming is more likely to involve auditory imagery rather than visual, showing that not all human daydreaming uses visual images.

Entertainment provides yet more opportunities for active play and for daydreaming. Becoming hooked on a particular form of entertainment most often results in actively watching, smiling, laughing, swinging one's body about and, above all, concentrating. Such concentration is usually accompanied by a host of moving mental images. Television, for example, provides specific contents for internal images - for example, watching a footballer on television often leads to images of self making the same moves. Television can provide the props for daydreaming both during and after a programme. Radio requires more imaginative concentration by self and as a result the learning may be even greater; as is the case with reading. In all entertainment circumstances, however, the content must be congenial to daydreaming. This happens when entertainment is 'positively' recognized by knowledge units (pre and developed). The content must also generate the mood conditions (see below) which make intense concentration almost, if not actually, automatic. Thus, successful entertainment most often reflects those things which activate daydreaming and play. As such it encompasses many of the things which are easy for humans to learn. That is, entertainment represent species typical human pre-knowledge and knowledge potentials and thus portends propensities for species typical behaviour.

In essence, daydreaming, play and entertainment are parts of the same perceptional and emotional-cognitive processes. During daydreaming, play and entertainment stimuli from certain objects, individuals and behaviours reduce internal agitation and/or set off heightened sensitivity and excitement. As a corollary, intense concentration results. During the process stimuli generated images interact with internal units of knowledge and, if they are not rejected, undergo an integration; as such they become new knowledge (and new potentials for motivating behaviour). In conjunction with processes of advanced cognition (see Chapter Six), the objects, individuals and behaviours of these images are often formulated into desires and fears (see next chapter), sometimes into fantasy targets.

Indulging in fantasies can be considered to be a rather intense form of daydreaming and/or desire formation. However, in fantasies the elaboration and intense charging of pure imagery goes somewhat further than with desires. Fantasies represent an aspect of human emotional-cognitive processing and a mental capacity which comes to depend more on mind (emotional-cognitive processing working on existing images) rather than experiences. As a result fantasies can become very unrealistic in terms of any chance for real achievement. However, fantasies are usually very easy to repress, or at least it is relatively easy to repress attempting to turn them into behaviour. Repression of fantasies, for example, can be through such relatively simple mechanisms as a sudden feeling of fear (if not terror) as to what the fulfilment of a fantasy might actually entail. Or it might merely be a fear of what its fulfilment would do to others (including one's own loved ones); or simply a realization that it is so impossible why worry about its non-fulfilment. On the other hand, individuals are bio-electrically motivated to achieve desires, and when desires (see next chapter) are turned into expectations (see Chapter Seven), humans set out to do so.

Nevertheless, fantasies represent an extreme development of human imagination and as such represent an important bio-electric dimension in the human learning process. A fan may fantasize about becoming a great film star, or of being seduced by one. They can know that it will not happen, but still learn a considerable amount about love, their own feelings and their own marriage aspirations from such fantasizing. A person may fantasize about becoming president of the United States, know that they have no chance, but still learn a great deal about American politics, politics generally, political power and the nature of the modern state while doing so. Moreover, fantasies can act as a source of potential pleasure during dreaming and daydreaming, providing a base for deferring gratification. That is, in so far as individuals get great pleasure from daydreaming about a fantasized future they are often motivated to face greater difficulties in the present than they would otherwise put up with.

In general the creation of knowledge units which are not congenial to daydreaming and fantasies are resisted. They are not easily stored in our long term memories or mentally experimented with. Any behaviours they suggest are not eagerly engaged in. If, therefore, we can discover universal patterns and contents of play, entertainment and daydreaming/fantasies we most likely will have discovered a great many of the universal patterns and contents of human desires and fears. These in turn will give us a good indication of the specific things which are for humans relatively easier or relatively harder to learn. But that is jumping ahead.

First we have to return to the dimension of *bio-electric charging* because in the end it is much more than a process of imagining which makes it more or less easy to learn things during play, dreams, daydreams, entertainment and fantasies. Bio-electric charging is, in fact, the dimension of emotional-cognitive processing which makes the human brain much more than a computer processing and storing species typical information. It is this dimension which makes humans have nightmares and panic attacks; it causes humans to *believe*, become *obsessed*, suffer depression, feel exhilarated; it is this which makes humans love and hate, fight and die, lie and cheat, sacrifice and worship. It is this, in other words, which motivates humans to be human.

Bio-Electric Charging

The argument so far has been that the human capacity to learn is very great, and even unpleasant things are learned (Humphrey, 1978). However, some things are considerably easier to learn than others (Shettleworth, 1972; Hinde and Stevenson-Hinde, 1973; Barash, 1982). This is because when stimulated in certain ways pre-knowledge activates a variety of bio-electrically generated *mood states* which make possible (or repress) the excitement and concentration necessary for certain spontaneous activities and for easy learning. In the process, mood states become linked to certain activities, objects and personas, becoming stored as species typical patterns of lust, love and jealousy, for example.

Various mood conditions, it is being suggested then, are set in motion when stimuli affect neural firing rates. Pre-knowledge units act as receptors for such stimuli. They also act as direction pointers with regard to setting in motion behaviour and in guiding stimuli to areas where additional knowledge units might be formed. Once formed, a number of knowledge units may themselves be provided with triggers for setting in motion the bio-electric actions which underpin mood conditions. These themselves, thus, become motivators of behaviour and of learning in so far as thinking, day-dreaming and reading, for example, about a particular knowledge unit can also trigger mood states - thus the notion of *charged memory*.

Mood states might usefully be classified into three major manifestations - those characterized as: 1) manic excitement, 2) depression (mental not neural) and 3) heightened sensitivity. Each of these conditions can greatly aid in the development of memory units (and motivate activities) but not all with the same efficacy. When the knowledge units which activate the bio-electric conditions we describe as the manic state are stimulated, we have motivations for intense but rather random engagement (probably of short duration) with a variety of stimuli. In the heightened sensitivity state we can anticipate motivations for intense, obsessional engagement with a limited number of particular stimuli. And in the depressive state, motivations relate to hyper, often disjointed and/or conflicting mental activity but at the same time to bio-electric repressors of physical activity. In all of these consciousness might seem to come and go, fade in and out, images replacing each other in a rather random pattern (Cf., Greenfield, 1995). We would also expect feedback linkages with knowledge units which activate anxiety and tension generating mechanisms in a number of the above combinations.

Heightened sensitivity conditions, then, seem much more likely to establish long term memory units than extreme depressive conditions, for example - although possibly resulting in obsessive specialization and/or commitment. For the most general of species typical learning, however, various combinations of mood states are likely to be most effective in encouraging or, conversely, repressing learning. The effectiveness of specific combinations, of course, will partially depend on what is being learned and the particular circumstances in which the learning is taking place. A mixture of heightened sensitivity and manic excitement seems the best bet for enhancing capabilities for awareness and concentration. Overall this mixture would provide the conditions most likely to result in long term human learning. As dashes of depressive mood conditions are added, this capacity will, we can speculate, become weaker and/or more diffuse. As a result we would experience less easy to maintain concentration and a vulnerability to boredom. Moving further along in this direction could result in clinical depression replacing boredom and it becoming almost impossible to retain anything new in memory. It might also be that here some of past memory would be blocked or lost (although some old material may be re-evaluated - both consciously and unconsciously – Cf., Taylor and Brown, 1988).

The existence of mood conditions is the reason that humans often experience and express powerful feelings. Humans can be motivated to expend a great deal of energy as a result of perceptions and memories being *charged* through mood conditions. This, as noted, is an aspect of the human mind as a motivational system which computer builders are a very long way from emulating and educationalists often dare not use. It is the capacity of emotional-cognitive mind which makes sex a preoccupation in all cultures, religious and political fanaticism common and drug pushers rich. It is a capacity which makes humans (at times) unafraid to risk death and/or be willing to kill each other. It motivates us to go on after major setbacks, to cling to each other with a furious jealousy, to worship heroes and gods with intense loyalty and to ignore the lack of empirical evidence for a purpose to our lives. In other words, the bio-electric charging processes generated during certain aspects of human specific emotional-cognitive processing make us, to a large extent, what we are.

General motivational direction in humans, then, is achieved through the interaction of mood conditions. Given that evolution works by adding to or restricting what is there, it can result in the evolution of neural processes which are often at odds (if not war) with each other. Such conflicting processes, it can be argued, often acts as the balancing mechanisms between relatively strong motivational forces on the one hand and repressive mechanisms on the other. In humans the strongest of these conflicts might be characterized as simultaneous tendencies for both addiction and addiction control. The generation of intense excitement and desire, over, say, sexual arousal followed by powerful senses of guilt - perhaps also embarrassment and shame - would be an example of this.

Based on pre-knowledge, we can relatively easily become addicted to, masturbation, sex, particular foods, tobacco, alcohol, drugs, specific others, heroes, gods, certain symbols, feelings of moral righteousness and a sense of obligation, for example. Such addictions, we can speculate, are often accompanied by feelings of *spiritual highs* (most likely some combinations of excitement and heightened sensitivity operating in a relatively intense fashion). Once developed, we can find it extremely difficult to shake addictions. But at the same time processes such as guilt and/or a fear of being rendered powerless by the hold of addictions over us, often accompany their development and restrain us (see below). In other words, humans have a considerable capacity for being bio-electrically punished for letting addictions become too strong. Humans are also rewarded for preventing excessive addictions from developing in the first place. This is often achieved through the extensive human capacity for deferring gratification (often through the development of guilt) and, very importantly, from the pleasure derived from fantasy daydreaming (see Chapter Seven).

Nevertheless, the processes of addiction and its control are so central to the phenomenon of bio-electrical charging during human learning and as part of the general human motivational 'system' that it is important to consider their possible bio-electric bases. To start, let us consider the usual experiences of a new baby. A new baby seems to bio-electrically respond to specific tactile stimulation, for example, such as being stroked, and to others gazing into its eyes. This stimulation undoubtedly activates protein synthesis in neural cells and electronic firing between specific neurons so that a number of more permanent connections between neurons are made. These, it can be suggested, transport and direct electronic signals to where they activate a certain combination of mood conditions, especially those of excitement/ heightened sensitivity.

Once certain circuits are established and sufficient connections are made with appropriate mood conditions, we can further speculate, anything other than stroking or gazing does not get through to excitement/heightened sensitivity receptors. It may be that the receptors have been made chemically "biased" (Cf., Greenfield, 1995) against non-stroking and non-gazing stimulation. This could even be the case when the receptors which activate these mood conditions are exposed to, for example, clothing, air, light . It could be that the complexity of the neural knowledge units generated by initial stroking and gazing require specific transmitters to complete all the circuits necessary to activate the excitement/heightened sensitivity route. Or, other signals might be diverted to tension/anxiety creating routes (expressed as agitation) because of some bias introduced into the receptor process. Whatever the exact mechanisms, the baby comes to 'need' tactile stimulation and eye contact. Nothing less will do in order to avoid tension/anxiety, to achieve biological homeostasis and general tranquillity.

At this stage, the bio-electric underpinning of addiction has been laid. Additional addiction generating stimulation and additional circuits can result in ever more specific neural pathways which require even more *specific* signals, such as, for example, that they must come from *specific* individuals. The significant point is that certain stimuli becomes *required* stimuli because substitutes either have no effect or are diverted to tension/anxiety generating mood conditions. It is a short step from here to postulating that a number of established circuits of this kind, *in the prolonged absence of the required stimuli*, generate their own anxiety directed triggers, and that these can only be overcome by relatively large doses of specific addicting stimuli. It is possible that as the above circuits become more elaborate it takes more of a particular set of stimuli to 'power through' to excitement/heightened sensitivity producing conditions - this phenomenon we might call, *'knowledge unit hunger'*. At this stage we have full-blown addiction. Heightened sensitivity derived addictions are often maintained for considerable periods of time. It is these, very possibly, which generate the spiritual highs so common among humans. In the end, it is not only the actual objects, behaviours and individuals which generate the images associated with addicting stimuli (and spiritual highs) but so too do memory images of these objects and behaviours. In this way, intense concentration and attention are directed to the objects, behaviours and individuals which make up the images associated with these conditions. (Dopamine and acetylcholine may be key neurotransmitters in these circumstances).

The evolutionary advantage of being able to focus intently on particular objects, individuals or behaviours, and to be motivated to expend considerable energy in mastering them, is clear. However, there is also a potential evolutionary danger. During the development of addictions an unusual amount of stimuli from objects of the addiction could generate sufficient dopamine or acetylcholine to create extreme highs (intoxication) and be counter-productive. The result could have been extreme excitement (mania) or extreme sensitivity (obsession) which, during evolution, led to disadvantageous reproductive and survival behaviours. Thus we can speculate that selection begin to favour the intervention of tension/anxiety connecting circuits at some point during the escalation of sensitivity and/or excitation to cut off further development of particular addictions. This might, for example, have been through activation of conflicting bio-electric tendencies (towards, for instance, both excitation and excessive sensitivity at the same time); or, it might have linked escalating sensitivity/excitation to depression generating circuits. Both connections would prevent fixation in the manic or obsessional areas - in evolutionary terms, the better to deal with, for example, unfaithful lovers and devious opponents.

However, given the add-on nature of evolution and in the absence of a grand design, tension/anxiety could act to induce a 'need' for more opiates and/or another combination of monoamines in order to counteract the tension/anxiety intervention. For example, an individual may desperately seek exciting activities to overcome impending depression only to experience even more depression as a result of neural confusion (see below) and/or letdown when their short term solutions are suddenly over. This in turn could motivate increased behaviours designed to stimulate excitement/heightened sensitivity, generating more tension/anxiety, and so it could go on. For example, drinking alcohol to escape depression can lead to excitement and heightened sensitivity for a while but then to more depression and a desire for more alcohol, and so on. In some cases striving to escape the cycle of depression and/or tension can become very intense.

This is what might be called the evolution of a *'frantic-feedback'* relationship among bio-electrics, images, thoughts and behaviours. Frantic-feedback relationships are those in which moods and reactions often dart in and out of extreme sensitivity, skipping now and again to excitement - from mania to depression, from care-free to obsessive, and back again. Individuals experiencing these are often on boarders, walking tightropes between sobriety and drunkenness, between sanity and insanity, and in their confusion continually generating tension/anxiety visceral and behavioural responses.

It is, however, in the circumstances of frantic-feedback relationships that the most dramatic characteristics of bio-electric addictions and their control can be observed. Frantic-feedbacks can lead to very high states of alertness and intense mental concentration. They can make an individual very good at masturbation, sex, sports, art, wheeling-and-dealing, drinking and war, for example. However, obsessive and continuous frantic-feedbacks can lead individuals into untold dangers. The above behaviour excessively motivated by frantic-feedbacks can lead to individuals becoming labelled as un-clean, wankers, sex fiends, jocks, artsie-fartsies, con-artists, drunks and blood thirsty war mongers. They can lead to an individual becoming rejected or shunned as being 'loony', unpredictable and uncontrolled.

While not completely detrimental to reproductive success, the obsessions above undoubtedly reduce it when compared to their more controlled forms. Therefore, we can suspect that selection pressure for controlling addictions, especially frantic-feedbacks, would have become relatively sophisticated. Let us consider how this might manifest itself in modern human life. Although we can become strongly addicted to our lovers, friends and heroes, from time to time we enter into bitter conflicts with them. We sometimes feel they are restricting us. We receive a variety of general disappointments from them. All this may lead to frantic-feedback attempts to regain the feelings of pure bliss involved in the original development of the addictions. But this can result in ever increasing disappointments. Slowly, or perhaps in jumps and starts, however, anger, a sense of betrayal and even depression or boredom build up, acting to weaken, if not break, our addictions.

But there seems to be another phenomenon at work as well in controlling addictions. As behaviour becomes increasingly frantic, more and more desperate, something can trigger new pathways, *suddenly discharging* (grounding) the stimulation. A physical example can be seen when a sharp induction of pain is used to overcome an intense itch. In this example, pain activates receptors which over-ride a frantic-feedback and for a time direct impulses to pleasure centres. In a more automatic example, individuals often go into a sudden state of startled relief, which includes a considerable improvement in concentration, as a result of just being spared a major disaster. A drug or emotional *overdose* (for example, too much alcohol or anger) can also suddenly dissipate, for a time at least, some frantic-feedbacks. In a social example, particular events can trigger a sudden blast of fear which quickly induces controlled behaviour, as for example usually happens when an individual just barely escapes being caught in an adulterous escapade. Sometimes, discharges almost instantly achieve a state of semi-exhausted exhilaration, if not tranquillity, as in having an orgasm climaxing sexual intercourse.

Overdoses and orgasms act as bio-electric overrides and/or discharge mechanisms preventing many frantic-feedback induced addictions from going to the point of generating continuously obsessional behaviour. A failure to achieve cut-off can cause an individual to experience, literally, a nervous breakdown; over a slightly longer term, individuals may develop clinical depresssion as a result of extended, uncontrolled frantic-feedbacks. While these control mechanisms are not especially favourable to reproductive or survival success, they do cause us to pull back, and sometimes re-evaluate our desires and expectations. In general, however, it can be speculated that selection resulted in the point of cut-off or over-ride being be at a point conducive to survival and reproductive success, other things being equal. Besides the examples of sexual climaxes and anger, other less dramatic examples include, the feeling of sudden and almost complete tranquillity when finding something extremely important relatively soon after discovering it missing; or the feeling following a period of stress, anxiety and panic about being unable to do something important when a sudden, unexpected, solution appears; or the feeling when alone and it is dark and scary and someone known comes by; or the feeling following a period of reflection, having concluded that self is useless, and then receiving believable praise from someone important, and so on.

In these experiences, we are motivated to search, discover a solution, be especially alert and to reflect on self via a frantic-feedback. But if this condition lasts too long we begin to suffer its bio-electric consequences (if not drive those around us to distraction, risking being rejected and/or being physically constrained). In these examples a 'positive' outcome triggers a relatively sudden dissipation of the frantic-feedback. What is also significant is that the events, objects, behaviours and individuals involved in the images related to the frantic-feedbacks and their sudden resolution have 'positive' lasting impacts on long term memory, future dreams, daydreams and future philosophical contemplation.

Such experiences often teach us that positive outcomes are always possible; as a result they tend to prevent us from, for example, taking loss of something as evidence that we must spend all our time obsessively guarding our possessions regardless of their value; they teach us to have hope. With hope come trusts; so even when something lost has been valuable, immediate feelings of anger directed at imaginary 'villains' may be sufficient to distract us from turning the loss into an obsessive guarding of possessions and/or total distrust of the whole world.. This may prevent the long term development of more general tendencies to anxiety/tension depression as over-riding dimensions of personality.

Addiction and its control are most likely to take place largely in limbic brain areas such as the pons, especially in the locus coeruleus, and the substantia nigra. These areas are typified by the development of single source/ divergent circuitry and by the production of the neurotransmitters acetylcholine, norepinephrine and dopamine. These areas are commonly known as the pleasure centres of the brain and are also involved in the mediation of stress, anxiety, arousal and depression. They are well connected to the neocortex where feelings, moods and information are stored as long term memory. Although endorphins are probably not directly related to pleasure as such (Bloom and Lazerson, 1988), receptors for them exist in abundance in the limbic system and may mediate (reduce) feelings of tension/anxiety. They would thus aid in relieving the pain of insufficient stimulation (Greenfield, 1995) and as such, perhaps, help provide for the dissipation of frantic-feedbacks, if not for the development of deferred gratification and/or the conscious control of emotional-cognitive processes generally.

SUMMARY AND CONCLUSIONS

The essence of the biology of human emotional-cognitive processes is that stimuli - internal, including genetic messages, and external - affect the production and firing rate of neural transmitters and the production and behaviour of certain hormones. Through this process neural linkages (networks or knowledge units) are generated and also bio-electrically charged via mood states. These knowledge units thus contain motivational information. During these processes, humans very easily develop bio-electric addictions to objects, persons and activities; however, a number of discharge and over-ride, orgasmic and/or tranquillity mechanisms, and a number of perceived danger feelings, have evolved which usually keep these under control. In the process motivation is provided for some species typical behaviours directly but, as we shall see, more often for the learning of more balanced species typical desires and fears.

As a result of all this, guidelines for stored energy are directed in a manner which makes it appear that an organism can defy the laws of thermodynamics. To use a non-human example, if I kick a ball of a certain shape and weight with a given force, it travels a specific distance according to known laws of physics. However, if I kick a dog of a similar shape and weight with an equal force, it turns around and bites me (this example was given to me by Alan Rowan). The dog has not just responded according to the laws of physics which apply to inorganic objects but has received the stimuli, evaluated it and drawn on a store of knowledge and energy guidelines (internal knowledge units) in order to overcome the laws of physics as they would apply to inorganic compounds. As a result the dog has responded in a much more, from its point of view, appropriate manner.

The above conceptualization is greatly influenced by current models from the physical sciences and from information theory. Indeed, models of the human mind have always been copied from currently known physical equivalents and it is just because of this fact that we have been able to understand the human mind as something more than a mystical seat of human spirituality. Every discipline reaches a point where it must be linked to concepts developed at another, adjacent, level of analysis if it is to move forward. This is currently true of the study of the human mind, behaviour and culture, and it is to the fields of micro-physics and electronics that we must eventually turn in order to find at least some of the insights needed to further our understanding of human nature. Indeed, it is interesting that, in the true spirit of consilience, scientists in micro-electronics are looking to the human mind and to organic processes and bio-electronics for hints as to how to move forward in developing future generations of computers. And it is clear that they have a very long way to go because the sophistication of knowledge units and knowledge unit potentials in the human mind, along with the powerful bio-electric charging that these often receive, are still well in advance of anything electronic engineers have so far been able to design.

The 'biology of emotional-cognitive processes' presented here is based on current understanding of neural physiology and not simply on a theoretically postulated black box. And, very importantly, at the same time it can provide a basis for linking genes and behaviour. Firstly, it allows for the developing of a model in which perception and consciousness are guided by pre-knowledge (certain genetically pre-programmed protein actions and/or neural processes, 'structures' and/or pathways). Secondly, it is easy to see how these processes are vulnerable to influence from external energy sources (stimuli) and from the release of trapped internal information (memory). And, thirdly, the potential for bio-electric charging (mood influencing) of knowledge units during neural activity makes them prime candidates not only for being the motivating force behind much of human behaviour but also for providing virtually all the psychological characteristics we associate with being human.

It is important to note that the R-Complex and limbic system of the brain, in which most of these processes discussed above either take place or are co-ordinated, enlarged as much during human evolution as did the neocortex which deals with more analytical processes (Geschwind, 1979; Crick, 1979; MacLean, 1978; Sagan, 1977; Barton and Dunbar, 1997). And, equally, the neo-cortex is deeply involved in storing charged knowledge units and in making associations between mood states and is thus as much involved with emotional-cognitive processing as with more advanced analytical processes. Whatever the details, human addictions, orgasmic experiences and perceived danger feelings become embellished in dreams, daydreams, fantasies, gossip, thoughts and theories. In the process emotional-cognitive processes become something not so easily describable in biomechanical terms. At this stage we have arrived at what might be called the 'psychology proper' of human emotional-cognitive processing.

CHAPTER FIVE
THE PSYCHOLOGY OF HUMAN EMOTIONAL-COGNITIVE PROCESSES

The psychology of human emotional-cognitive processes involves the turning of charged knowledge units into relatively specific, that is focused, species typical 1) emotional-cognitive motivators and 2) desires and fears. This involves a greater degree of cognitive interaction and abstract thought than has so far been considered. There is more combining and recombining of knowledge units and the creating of units to represent other units (symbolizing). It is likely that this dimension of humanness evolved as *conscious awareness* and *self-analysis* began to give reproductive advantage (Crook, 1980; Jolly, 1988; Buss, 1994). Nevertheless, psychology as opposed to physiology is a matter of degree rather than kind. The notion of a clear demarcation between them is more an academic demarcation than one which represents something real in nature. This should be clear as we consider species typical emotional-cognitive motivators and desires and fears more closely.

SPECIES TYPICAL EMOTIONAL-COGNITIVE MOTIVATORS

We can usefully start our examination of these processes by trying to identify those traditionally designated emotions (as emotional-cognitive processes) which might have more scientific meaning than others. In this regard the intention will be to examine a number of 'emotions' in terms of their possible bio-electric, behavioural, motivational, addicting, social and/or philosophical manifestations. Sociological, psychological, anthropological, historical, fictional and visual literature suggest a range of universal human emotional-cognitive processes which, in traditional terminology, include at least: pain, anxiety, shame, greed, hate, pleasure, surprise, disgust, guilt, envy, joy, depression, panic, jealousy, loneliness, empathy, excitement, fear, anger, sadness, interest, embarrassment, grief, elation, curiosity, happiness, lust and love.

Many of the activities above can best be understood as bio-electric processes at neural, cellular and visceral levels. Others, on the other hand, more appropriately represent philosophical concepts used to make ethical and moral judgments about a given state of humanity. Some, however, are sufficiently broad for both extremes to apply. For example, the concept of pain can be used to designate, or represent: 1) certain chemical reactions, 2) specific types of neural transmissions, 3) identifiable muscular reactions, 4) specific visceral and facial responses, 5) experiences of discomfort, 6) specific mood conditions, e.g., agitation/ anxiety, 7) certain mental states, e. g., clinical depression, 8) unpleasant social experiences, 9) feelings generated by specific thoughts/memories, 10) feelings set in motion by the sight or memory of certain individuals, 11) embarrassing situations, 12) a vision of cruelty in action, 13) disgust felt in terms of perceived immoral acts and 14) a means for demonstrating endurance, of demonstrating self-actualization.

What pain represents in people's minds clearly can exist all along a continuum of processes between genes and philosophy, although the emphasis can vary considerably from individual to individual. Differences of meaning become especially obvious when we compare those who interpret human motivations in purely physical terms (genes and 'brain wiring') with those who hold a humanistic view (development of self-understanding). Pain for many biologists is nothing more than the genes' way of making an organism avoid dangers to those same genes. For some humanists, on the other hand, pain has little significance separate from the meaning given to it by reflective humans (Trigg, 1970) - especially those striving for self-actualization.Although neither of these perspectives tells us what pain *is*, they both open up important areas where a notion of pain does have significance for explaining human behaviour. In a sense they are both right. The problem is to conceptualize the bio-electric processes associated with pain in such a way that it is possible to link them to the psychological processes equated with pain, and these, in turn to behaviours and social processes which hurt (physically and psychologically). Then we would be on the way to developing a theory of human motivation generally, while, at the same time, being able to uncover some of the actual mechanisms/processes through which certain types of human social relations/ideas motivate humans to avoid danger to their genes (indeed, even to perpetuate their genes).

Other 'emotions' too exist at several locations along a continuum from genes to philosophy, and certainly none at only one. For instance, the same ambiguity (and problem of definition) also applies to the phenomena of: fear, disgust, pleasure, happiness, excitement and panic. Fear, as a case in point, is probably best described as an emotional-cognitive processing of danger reactions. Danger reactions, such as fright, disgust, tension and anxiety, for example, are most likely derived from excess excitation and/or particular conflicts of neural excitation and inhibition (see below) at the bio-electric level (bio-electric pain). These can simply result in the general feelings of danger which self in most circumstances struggles to control, but, at the level of behaviour, can result in panic, anger, aggression, violence, avoidance, flight, clinical depression and so on.

When the above bio-electric processes are linked to specific cognitive orderings (that is, when specific *images*, for example, become bio-electrically charged) the results can be considered 'human fears'. At this level fears are often manifest in the feelings and reactions we describe as disgust, shame, embarrassment, jealousy and guilt. And, when dreamed, daydreamed and thought about, human fears can take the form of stereotyped images of the imminent dangers thought to be presented by 'the anti-god', witches or foreigners, for example. The concept of fear thus provides valuable insights and questions about a number of neural processes, behavioural responses and even how humans conceptualize danger inspired behaviours, but renders an essentialist definition of fear confusing, if not impossible. If pleasure and happiness mean a general heightened sensitivity, tranquillity and controlled excitement, they coincide with some of the bio-electric processes discussed above. If they mean an absence of something like shame, guilt or anxiety, they coincide with the above processes interacting with more advanced emotional-cognitive activities, and if they represent a sense of security based on a feeling of 'togetherness with God', they are philosophical concepts derived from mood highs generated from images of supernatural purposes and beings.

Some emotional-cognitive processes are largely bio-electric processes manifest in involuntary visceral and behavioural activities. Excitement, elation, sadness, surprise, interest and curiosity represent such processes. Each of these words describes certain physical (often visceral) reactions to particuar combinations of neural excitation and/or inhibition which result in tendencies to heightened sensitivity and/or depression in terms of mood states. Despite their scientific vagueness, however, even these, as social concepts, can be given meaning in terms of human social behaviour and human values. If humans feel it necessary, these can even be turned into both causative (deterministic) and philosophical 'essences'. Greed, joy, interest (perhaps), grief and loneliness seem to be somewhat more analytical (involving more advanced cognition) or even contemplative/philosophical concepts. Greed is a social description of persons or behaviours (including those of self) which appear dangerous to others because they seem to want more than their 'fair-share' of things. Joy may start from feelings of elation (a bio-electric phenomenon) but usually also includes an analysis of the elation and excitement in terms of rationalistic explanations given by both self and others as to its causes. The condition we term loneliness may arise from a mild depression generated by daydreaming about and/or contemplating images of being rejected, isolated, not loved. But this condition, and its possible causes, lends itself to analysis, by self and others, which gives the notion specificity.

Most of the emotional-cognitive processes discussed above have some meaning but lack the precision needed for developing a model of human nature which links genes and behaviour. Their use as conceptual tools is clearly limited but nevertheless their descriptive value is often very useful. Therefore, throughout this book, they will be used from time to time for purposes of un-encumbered communication in the hope that the context will make clear the meaning intended. The reader should remember, however, that these processes as emotions or as 'elements of cognition' are not being given the status of essential 'things' with causative powers in and of themselves. Nor are they to be considered as unique human capacities with specific above-nature characteristics. Rather they will be treated as representing a variety of processes operating at various 'points', if not at all points, along a hypothetical continuum running from biological processes to philosophical contemplation.

Lust, love, jealousy, hate, envy, empathy, shame, embarrassment, guilt and depression do have a more direct specificity which is relevant for the model being developed here; it can be argued that they more closely approximate what we can refer to as *species typical* emotional-cognitive processes. Interacting together - and with advanced cognition - they represent the key motivational processes between genes and species typical patterns of behaviour. They, therefore, require more detailed elaboration in the light of our discussion of the biology of human emotional-cognitive processes. From this vantage point it is possible to begin to see how these are more than bio-electric processes; that they clearly begin to demonstrate a psychology of motivation.

Lust

Lust involves very highly charged bio-electric processes which easily generate addictions to certain body parts, expressions, behaviours and the general shape of certain persons and objects. Images of these are stored as knowledge units (often as visual images). When stimuli activate these units, excitement and tension can reach levels of intensity which run out of conscious control. Heads turn quickly to fix vision intently on objects of lust, vivid daydreaming commences and elaborate fantasies are spontaneously generated. Neural, visceral, fantasy, memory and behavioural activities enter into frantic-feedback relationships - muscles tense, mouths water, penises rise, vaginal fluids flow. We feel compelled to grab and hold, squeeze and hug, almost to the point of doing physical harm to, the objects which set lust reactions in motion.

Pre-knowledge units and unit potentials in our brains largely determine the types of lust knowledge units which are possible. Pre-knowledge units generally include the basic template for the objects, individuals, behaviours and physical characteristics which stimulate lust. The details, however, are filled in by experiences. Experiences can also change the focus of lust. Familiarity and boredom, for example, can act as an over-ride from focusing lust on a specific body part, face or person, for longer than is reproductively advantageous. This diverts attention to other activators of lust, possibly increasing reproductive success. Nevertheless, lust for certain objects, characteristics and persons is very easy to learn, if not become addicted to, and in many circumstances lust reactions can be very difficult to repress. Indeed, lust knowledge units often develop the capacity to generate tension and anxiety when lust stimuli are not forthcoming (knowledge unit hunger). Such deprivation additionally generates anxiety and tension which can reach pitches of intensity which often lead to seemingly out of control behaviour.

Lust reactions are activated by signals of various combinations of submission, affection, sexual receptivity, mastery and general childlike presentations of self. That is, lust knowledge units contain information which generally: attracts males to females, females to males, infants to parents, parents to infants, people to heroes and heroines and to objects of beauty. (A search for specific objects, behaviours and characteristics which generate lust reactions will have to become an important part of our science of humans). However, as is typical of other powerful emotional-cognitive processes, much of human lust is, most of the time, repressed and/or controlled through the actions of a number of emotional-cognitive regulators, not the least of which has been the evolution of love.

Love

Love, we can postulate, develops as the frantic-feedbacks and intense excitements characteristic of lust knowledge units (images) are integrated with, developed into, or replaced by, knowledge units also charged with considerable heightened sensitivity and opiate generated tranquillity states. Strong wishes to hold, hug and squeeze objects of lust are tempered by additional wishes to groom and caress them. Gentle holding, hugging and grooming behaviours feedback to further heighten sensitivity and very likely to produce additional endorphins. These kill pain while norepinephrines fire rapidly and weariness disappears. Adult lovers talk into the night, they laugh, they giggle, they play, they invest trivial places, occasions and events with special, spiritual, significance (van Sommers, 1988; Pope, 1980; Eibl-Eibesfeldt, 1971; Peele and Brodsky, 1977; Money, 1980; Buss, 1994).

The wish to hold and hug is, thus, soon accompanied by spiritual highs generated simply by the proximity of love objects with self. Humans have strong feelings of warmth and safety when in the company of love objects and want to be with them for long periods of time. Humans wish to be protected by, and to protect, love objects. In general, elation and joy flow through lovers' bodies when their love objects come into view and a sense of spiritual well being develops in their presence. This often increases and/or becomes more secure with continued association with them. Fantasized images of love objects become essential parts of holistic images of happy, fulfiled self. And so lovers very often automatically treat love objects as personal property.

This means, however, that individuals can be extremely vulnerable to rejection, to loss of a love object (which is like having some part of themselves being taken from them). Loss of love can be a shattering blow to fantasies of self wonderfully happy in a fantasized future. Such fantasies are often replaced with images of self totally alone. This is because, owing to the specificity of the bio-electric energy that has been invested, there are no quick, easy replacements when lovers withdraw love. Love can also become restricting (engulfing) of others. Strong love feelings not perfectly reciprocated can, in effect, lead to a great deal of mental, personal and social conflict. As a result, love, and the other emotional-cognitive processes it generates (see below), can be expensive: it can use up a lot of energy and can even detract from relatively more successful reproduction and survival practices if indulged in to excess. Therefore, selection seems to have restricted us to the extent that love feelings are sparingly used; only a few individuals and/or objects will receive the full attention of love feelings during the lifetime of a given individual (Cf., Freud, 1957; Buss, 1994).

The specific bio-electrics of human love would be something like this. When love objects express love attention to self (actual and in images) the result is the development of relatively closed circuits with neural connections to excitement and heightened sensitivity generating mood areas. Stimuli from other objects, including those which normally would generate lust, are progressively excluded, if not diverted to tension/anxiety generating mood conditions. This would represent a more charged example of the closed circuit circumstances which tend to prevent the learning of a second language after a first one has become firmly established. As such it would go some way to explaining the often exclusive focus and tenacity of love.

At the same time, humans have evolved a number of emotional-cognitive mechanisms which prevent them from jumping headlong into unevaluated love adventures (for example, coyness, shyness, embarrassment, desires to be independent, to be 'strong'). Also, humans have evolved powerful feelings which motivate them to guard ferociously the love relationships they eventually decide to risk all with. One major example of this is the considerable human capacity for jealousy.

Jealousy

In order to separate jealousy from emotional-cognitive processes such as envy, anger, hate, and so on, it is possible to consider it as a powerful and deep anger - sometimes expressing itself as clinical depression - generated by a loss, or feared loss, of a love object (Van Sommers, 1988; also see Wilson and Daly, 1992; Buss, 1994). The bio-electrics of jealously might be something like this. If images of love objects expressing love attention to self progressively generate relatively closed circuits, these circuits could also have divergent connections which direct stimuli to tension/anxiety mood areas when love objects pay considerable and intimate attention to *new* or *additional* objects. Given enough stimulation and charging these units can, of course, become stored as memory. Building upon this basic love/jealousy circuitry, additional diverging circuits (sub-knowledge units – images) are, we can speculate, also generated. These would include images: in which self is abandoned by the love object, in which a love object is happily ensconced with some other and so on (the importance of visual imagery here seems very clear). These too can become stored as memory, so when themselves stimulated can activate the main units. In other words, once established jealousy can have internal memory sources. Thus, through day-dreaming and fantasies alone, images of love objects in the wrong proximity can lead to feelings of rejection, humiliation, if not destruction of self generally.

Jealousy images can be activated by a number of personal and/or social events, such as love objects cuckolding, rejecting, abandoning, humiliating and generally scorning self. It is likely that given individuals have a greater or lesser tendency to feelings of jealousy. This may be based on differential inheritances of personal capacities for pre-knowledge and/or developmental love/jealousy circuitry, or by a variety of different childhood experiences. For example, parents might establish very close, stroking, grooming, hugging, cuddling, gazing intimacies with offspring, but later distance themselves in order to pay more attention to newly arrived offspring and/or to give more attention to each other - or to a new mate. The relatively sudden withdrawal of attention may be very fertile grounds for the growth of jealousy knowledge unit potentials, potentials which have effects at the time but also become stored and 'ready' for new content later, as such making individuals prone to fits of jealousy in later life. This example also shows that it most often is not pathological behaviours which makes individuals prone to jealousy, but rather normal reproductive behaviours; certainly jealousy serves reproductive ends in post childhood life.

Whatever the exact causes and behavioural consequences, it seems very likely that early jealousy experiences can result in images of self being abandoned and these can remain as charged memories (sometimes as unconscious memories). As a result, specific experiences of rejection are not necessary in later life for an individual to become jealous. All that is needed is for some current event to trigger pre-existing jealousy knowledge units. Moreover, it is not difficult in this state for imagination to create a worst case scenario about where current events might end up due to an individual's own imagined vulnerability. Indeed, it is common for humans to be jealous of the past lives of their mates, of ex-lovers, of ex-wives/husbands and so forth (Van Sommers, 1988). Here jealousy is directed at individuals from the past, whom they have never met and who may even be well out of any imaginable personal future. And, definitely, it is images of trusted partners divulging to some other the intimate secrets developed conjointly with self which generate the greatest degree of jealous anger and depression (Van Sommers, 1988). In other words, *images* of individuals granting intimate attention to others, and/ or abandoning or not being with self, *even in the past, future and out of sight*, activate the psychological feelings and bio-electric bases of jealousy just as much, if not more, than real events.

The evolutionary significance of jealousy in the above sense is that it motivates individuals to a number of behaviours which tend to promote re-productive success, given the likely scenario of human evolution in which monogamous matings became increasingly common (see note four, p., 234). Jealousy clearly motivates individuals to continually guard their love objects against loss to outside individuals. It also motivates humans to *evaluate* very carefully the potential loyalty of individuals before love relationships are en-tered into. Moreover, jealousy, in moderation, tends to show a partner that one really cares and that self will risk a great deal in defence of the sexual and love relationship established.

On the other hand, humans do not like excessively jealous others be-cause they threaten self's independence, place great demands on self and can be generally disruptive. Excessively jealous individuals often feel extremely angry, hurt and aggrieved, while others perceive them as out of control, over-possessive, not to be trusted, certainly not loved. For jealous people these types of real or perceived rejection are prime conditions for the de-velopment of bio-electrically painful feelings of guilt and extreme depress-ion (see below). Nevertheless, jealousy remains one of our prime emotional-cognitive motivators, universal and extremely significant, not only in moti-vating individual behaviour but also for providing a link between individual bio-electrics and social behaviour.

> "Jealousy goes beyond a personal experience. It is a potent social force, like a sleeping serpent whose capacity to uncoil and sting affects all our emoti-onal plans and ultimately shapes the whole geography of our social arrange-ments as well as our affectionate voyaging." (Van Sommers, 1988, p., 195).

Hate

Hate, like jealousy, is a relatively complex emotional-cognitive process. It includes a degree of cognitive ordering and specificity not found in simple aggressive or flight reactions. Images of certain combinations of particular individuals, objects and/or behaviours make up knowledge units which, when stimulated, activate tension and anxiety linked circuitry. Frantic- feed-backs stabilize at relatively high levels of intensity; that is, they do not cause completely random activity but they do not easily discharge either. Again, the significance of images is very clear in that it is often much easier to hate an image of the other, for example, than to confront them.

Hate images are derived from experiences of receiving physical harm, being abandoned, feeling jealous, being rejected, being bested, being humiliated and feeling frustrated by social others. As a result of hate we often become especially vigilant in our social behaviour; we spend a lot of time trying to order our behaviour so that those we hate will not humiliate or hurt us. As a result of hate we can become incredibly introspective regarding our own motives. In so far as reproductive success is measured in terms of individual successes in competition with others, the value of hate as a motivator of evaluation and vigilance needs little explanation. However, hate can easily become obsessional and, as such, can become a counter-productive addiction.

Envy

Envy is the emotional-cognitive condition we feel when we see others having something we wish we had, or indeed feel we should have (Schoeck, 1970; Foster, 1972; Willhoite, 1981; Van Sommers, 1988). It is likely that the bio-electrics of envy are very similar to those of jealousy and hate. Images of self being happy come to include the possession of objects, persons and behaviours possessed and experienced by others. It is common that others are blamed for self not possessing these. "Envy usually does not act: it complains, it vilifies, it gossips maliciously, it moralizes, it rationalzses" (Carroll, 1985, p., 25). There is no doubt that envy is a universal and a very powerful force (above references). As such envy motivates a number of behaviours which provide quite complicated social behaviour.

This has come about because envy, like jealousy, motivates a number of behaviours which undoubtedly provided (and still provide) reproductive success. For example, envy is likely caused and fuelled by fear of and actual failure (Carroll, 1985), as well as by fear of and actual loss; in so far as envy motivates us to avoid these, its potential 'positive' reproductive consequences are clear. But there is more to it than this. People are not only universally envious but clearly mistrust those who *show envy*. This is because envy is a clear signal from those who show it that they are not satisfied with the 'rights' they have been granted; they might be motivated to vengeance and to grab and cheat. So, *a fear of envy* most likely also evolved as a means of identifying and controlling potential cheaters in social life (Willhoite, 1981). Certainly, envious individuals are considered socially dangerous and are reprimanded, stigmatized and sanctions are exercised against them.

We can speculate, however, that human fears of envy acted as selection pressures for individuals to be *afraid of showing envy* rather than against envy itself. That is, because envy has undoubtedly given reproductive success it is unlikely that selection would have eliminated it. But selection certainly would have favoured those who were able to *conceal* their envy. Such individuals would have had envy to motivate them to achieve (including at the expense of others) those things which provided reproductive success but an ability to avoid sanctions because others were deceived into thinking they were, say, altruistic. So, for example, a man might give a gift to a child because secretly he intends to seduce his mother (at the expense of his father). 'Uncle' is a nice generous guy, daddy a tight wad because he does not give his kid presents, daddy seem petty if he rejects the gift for his boy - mommy might just take some notice of 'uncle'.

Another selection force for concealing envy would have been something like the following. In so far as showing envy represents weakness (as evidence of failure, loss, achieving less than others, and so on), the ability to avoid others taking advantage of self in a weakened condition is enhanced by concealing envy. This is evidenced by the fact that we have clearly evolved a capacity to feel embarrassed, guilty and shameful when our envy shows itself. "We often pride ourselves on even the most criminal passions, but envy is a timid and shamefaced passion we never dare acknowledge" (La Rochefoucauld, 1959 [1665], p., 38). The human nature process for concealing envy can include emotional-cognitive evaluations of self whereby images of self as being strong, self-reliant, not dependent on the success or attention of others come to predominate - that is, through the development of what we call personal confidence. Expressions of humility and/or modesty are also strategies adopted to prevent feelings of envy being directed towards self, and can probably give secret feelings of personal superiority and may even gain prestige for self.

> "Moderation is a dread of incurring that envy and contempt which people drunk with their own success deservedly bring upon themselves; it is a pointless display of our own greatness of soul. In a word the moderation of men at the height of success is a desire to appear even greater than their destiny (La Rochefoucauld, 1959 [1665], p., 37).

Envy, however, is, as noted, a universal and very powerful and its control is far from having been achieved by unconscious and/or 'rational' constraints. In reality loss and failure are always possible, making envy and its control a continuous emotional issue. In order to control envy an individual must maintain constant vigilance over his or her own behaviour and responses. In order to achieve the most from others it is necessary to achieve a balance between humility and personal strength. Humility must not go to the point of showing personal weakness (humiliation) and strength must not go to the point of excessive domination (arrogance). To achieve this to maximum advantage in a variety of situations it is helpful to be able to read the potential envy of others in advance, in order not to arouse it to vindictive or defensive action. Being able to put self in the emotional-cognitive mind of another (Cf., Goldstein and Michaels, 1985) is a means of doing so. We call this empathy. Empathy may well be, in part at least, an evolutionary result of a fear of showing envy.

Empathy

Empathy is the ability to feel as someone else feels. Current research on autism clearly shows how important this ability is in 'being human' (Cf., Baron-Cohen, 1995). Empathy makes it possible to know and feel ahead of time if someone is likely to be angry or happy about the behaviour of self. Above all, a capacity for empathy makes it possible to avoid the danger of rejection. It makes it possible for a whole variety of behaviours to be altered before they reach the point of driving parents, potential lovers and friends away (or, on the other hand, of leading to engulfment). Empathy allows an individual to know if his actions are likely to receive positive attention or, conversely, condemnation. Empathy also makes it easier to deceive others and to spot and guard against deception (Cf., Hauser, 1997; Whiten, 1997).

During the development of empathy individuals often emotionally identify with those whom they perceive to feel and think like themselves, those who want to play the same games as themselves. Bio-electrically, as individuals build images of such others, images of the other come to be charged with heightened sensitivity mood states. It is also likely that the heightened sensitivity 'high' attached to the image of other is further activated by images of self interacting with them. In so far as this becomes part of self-image, other becomes part of self. In so far as there is confirmation of this linkage a degree of safety through predictability in a context of 'positive' emotional-cognitive charging is created.

Feelings of empathy result in a considerable amount of altruistic behaviour being extended to those with whom one has empathy because, at least to some extent, the welfare of other has been incorporated into one's own self-image. Self desires continual recognition of other and self expects unquestioned reciprocity (because self would do it for other). One result, however, is that most other humans are either ignored or stereotyped as being very unpredictable, and thus potentially dangerous. Once 'positive' identification takes place individuals become friends, lovers, caring parents and/or devoted offspring. At this point almost all behaviour among them comes to be more or less acceptable and forgivable. Danger to offspring, parents, friends or lovers becomes danger to self, and harm to them becomes hurt to self. Trust is as near total as it can become. When highly charged (perhaps heightened sensitivity is spiked with a dash of excitement) the bonds established in the above relationships seem to have supra-individual, if not supranatural, qualities. These have been described as Love, Dedication, Honour, and Fidelity, for example. As a state of being the concept of 'romantic love' represents fantasized versions - greatly elaborated - of hoped-for empathetic love encounters and relationships. Each partner is expected to fully know, understand and, above all, appreciate, the other; indeed, a complete fusing of self and a specific other is often expected. To a great extent the notion of romantic love combines the bio-electric forces of love and empathy into a social/ philosophical expression.

Empathy is very often extended to stereotyped groups so that whole groups, or at least images of them, become, if not self-like, at least acceptable allies of self. In feeling that self is in tune with a group, self, in a circular fashion, is motivated to conform to the norms of the group in order to substantiate this feeling. It is from this that we get senses of sisterhood and brotherhood. Feelings of empathy, thus, help make an individual's immediate social environment understandable and relatively safe through the support it gives for the generation of and adherence to shared norms and mutual obligations. While love might not be extended to individual members of such groups, feelings of jealousy, hate and envy among members are usually greatly reduced, or at least compensated for. Moreover, feelings similar to those generated in bio-electric love are sometimes extended to the shared objects, ends and symbols of such groups – for example, to shared gods, icons, music, heroes, leaders and flags. To allow empathy to operate in human relationships, however, requires considerable self-control and often the demonstration of devotion to both others and symbols of what the others represent. A major emotional-cognitive mechanism for maintaining and enforcing such feelings has been the evolution of the considerable human capacity for guilt.

Guilt

In one sense guilt is something we wish others to feel when they transgress against us in order that they might suffer for their transgressions and not transgress again. This likely derives from an easily developed human desire for vengeance (see Chapter Seven). At the same time, however, our fear of reprisals makes it very easy for us to learn to feel anxiety about doing things we and significant others feel are disapproved of. This form of guilt is dependent on a relatively well developed capacity for empathy and reason and can be called moral guilt. Moral guilt also derives from our fear of failure, from feelings of envy and our fear that this envy will show. In a more important sense, however, effective guilt derives less from emotional-cognitive processes interacting with conscious reason than from much earlier developed knowledge units linked to 'negative' charges. In this sense guilt is felt as internalized pain when even *contemplating* certain actions regarding others. Its learning is largely unconscious (most of it probably takes place at a very young age), and it is often sufficient on its own to restrict threatening behaviour towards others. As such, guilt acts as

> ". . .man's sixth sense, the grey eminence that commands him what to do, that divides the world into good and evil, that takes him out of the pure vanity of being concerned with his self and its pleasures and makes of him a man . . .Guilt is not merely the dark force to be kept in check, to be constrained and to be harnessed. It is our richest most hidden resource, the essence of humanness" (Carroll, 1985, 1, 4-5).

We learn both moral and internal guilt very easily and, as noted, the latter usually very early in life. Like oral language, internal guilt is, in fact, so easy to learn that we cannot help but acquire an appreciable degree of it. Before we learn about socially defined wrong doing, and even before we develop a fear of showing envy, there are, Carroll argues, two aspects of human pre-knowledge (my word) available for the very easy, natural, learning of internal guilt. They are 'primordial fears' (his words) of 1) powerlessness and 2) separation. These, he argues, lead to two major forms of guilt, "persecutory guilt" and "depressive guilt". In our terms, addictions to images (knowledge units) of feeling in control of life sustaining conditions - physical and mental - and/or for perpetual inclusion in safe havens, when not stimulated, generate intense danger reactions. These danger reactions, in turn, are stored as knowledge units which, when activated by stimuli which suggest a losing of control and/or separation (general loss, rejection, lack of affection, frustrations and so forth), activate combinations of mood conditions which we experience as pain/tension/anxiety.

However, these 'negative' units have the additional characteristic of directing the expression of this pain/tension/anxiety away from behavioural motivators for aggression against the emitting source of these stimuli. Instead they motivate substitute activities or the turning inward to an attack on self. Guilt which motivates substitute activities is what Carroll calls persecutory guilt and might, possibly, win back attention. Tension/pain/anxiety linked to these guilt knowledge units tends to be dis-chargeable by activating images of self doing 'good works', of being worthy of receiving 'positive' attention, for example.

In the second case, in bio-electric terms, guilt units direct tension/ anxiety generating stimuli inward, towards images of self, making self seem to be the *cause* of the pain generating stimuli. This is the case wherever the stimuli really comes from. This is depressive guilt. In depressive guilt, feelings of pain/tension/anxiety can only be relieved (discharged) by activating images of self desperately making reparations to the *perpetrators* of the danger stimuli. In other words, individuals blame themselves for wrong doings, even by others to themselves. As a result, individuals so affected are often motivated to withdraw from social interactions, sometimes through the bio-psychological mechanisms of depression (see below). In a less extreme example than depressive withdrawal, consider a child who says, ‘I can't do it, I am no good at *anything*' in response to criticism for a minor failure; or individuals who blame themselves for failed relationships when outsiders would put the blame on their partner.

The bio-electric mechanisms of guilt, then, generate intense anxiety, tension and behavioural depression by activating memory units which are highly charged with danger information. As such, feelings of guilt act as very powerfully charged spaces between self and others and between self and certain behaviours. From an evolutionary point of view they prevent us from directly attacking the sources of our protection (also the sources of our powerlessness) when we would otherwise be disposed to do so. A very small child, for example, can be very aggressive in trying to have its way but soon learns that too much aggression can lead to hostility and even separation from the care-giver. Throughout life, whenever we attempt to activate our drives for power, a remnant of an early fear of separation, or more current fears of rejection, exclusion and/or isolation, frequently pull us back.

Very often, during their development, feelings of guilt generate ambiguity and confusion between our tendencies to aggression and our desire for love. Questions arise: should I pay more, or less, attention to a particular person in order to win his/her love?; do I seem weak and ineffectual, or aggressive and arrogant?; am I acting as a hero, or a coward? This ambiguity can become highly charged, it can generate frantic-feedbacks which are potentially very destructive. But it may be this very process which often motivates us to reconsider our relationships, our boundaries and the binary divisions (see Chapter Six) we use to classify the world of things and people. Guilt generated by ambiguity may motivate us to create some new boundaries, behaviours and attitudes. Thus the sometimes expressed view that guilt is the driving force of civilization.

Overall, however, guilt can be very painful and it is generally accepted that humans use a number of avoidance techniques in order to stay clear of situations which activate their guilt mechanisms. Humans usually avoid circumstances in which they might have to put blame onto themselves: in particular avoiding, whenever possible, putting self on the line for judgment, risking loss of esteem, status and power and/or of having to operate in conditions in which they have little idea of what is required of them.

However, many of the above circumstances are unavoidable. This may lead to self-accusations in order to take the sting out of anticipated failure. At the same time, individuals may set about strengthening the barricades around self, further restricting their own behaviour as a result. On the other hand, individuals may turn outward, projecting self into a strong adherence to the rules, taboos and social boundaries which seem to them to eliminate, or at least control, loss, failure and/or ambiguity. Taboos, rules and boundaries provide a more or less formalized method of establishing *rights* so that encounters between individuals are less dangerous than they might otherwise be. The establishment of rights designates, ahead of time, the amounts of envy and aggression an individual can expect to get away with. Thus we can see that with envy, empathy and guilt we are approaching the emotional-cognitive processes which underpin human social organization, human hierarchies and human politics.

Embarrassment and Shame

Much of the time, then, human social encounters are potentially dangerous, both psychologically and socially. Self can easily be rejected, isolated and shunned and as a result, perhaps, be consumed with extensive depressive guilt. Given the reproductive disadvantages of the above we would expect that selection has provided us with a number of emotional-cognitive mechanisms which would identify and help us avoid and/or quickly retreat from interaction which might lead to them. Infants soon develop a tendency to *shyness* when in new situations, or are presented to people they are not familiar with. This seems largely to be an unconsciously learned response to novel circumstances and individuals.

With the development of self-awareness infants begin to sense that certain behaviours are inappropriate. They come to discern that being discovered will lead to punishment or ridicule; they come to realize that certain behaviour will get them less loving and more rejecting from those they are becoming addicted to. As a sense of self begins to incorporate a notion of self-value, behaviours which might lead to ridicule and/or rejection loom larger and larger. Neural connections – influenced by pre-knowledge – between knowledge units created from having committed behaviours which are likely to be ridiculed and rejected (or even mental images of possibly committing some of them in the future) and tension/anxiety centres, cause discomfort. Such conditions feedback on visceral and heart/blood functions so that there is a loss of homeostasis which most often results in a quick retreat, not only from such behaviours but also, often, from sight. In other words, situations which stimulate images of ourselves being rejected, humiliated, looked down upon, generally scorned and so on, generate almost immediately the mental and visceral changes which we usually describe as *embarrassment*.

Once we start to become preoccupied with our presentations of self -consciously or unconsciously - these tendencies increase. And when we have generated a set of taboos and boundaries - codes of right and wrong and of good and evil - to guide our behaviour to the end of avoiding dangers in social life we can further become subject to embarrassments if we violate them. This usually prevents us from going so far that severe guilt and depression are the result. Indeed, we soon develop concepts of *shame* which tell us which situations to avoid in order to prevent the physical discomforts of embarrassment. Shame might be thought of as being a half-way-house between embarrassment and guilt, or at least as an alternative which has social, rather than internally inspired controls on the behaviour of self. Shame usually is only felt when we know others have seen us misbehave, and we feel, therefore, that we will be rejected. Guilt, as we have seen, can go to the point of us blaming ourselves for even the 'negative' things that happen to us - even when we are not seen, recognized or in any way held accountable. Shame, on the other hand, can often be avoided by sinning in private, it is very difficult to hide from guilt; shame can be overcome with denial and honorable deeds in recompense; guilt is not so easy to shake.

Nevertheless, a sense of shame can have strong bio-electric connections, and does, in fact, share a number of characteristics with guilt - or at least some forms of guilt (Cf. Carroll, 1985). In some cultures concepts of shame are made the bases of many social rules and hierarchical evaluations, so that a great deal of social life is predicated on behaviours designed to save face, that is to avoid shame (Cf., Nisbett and Cohen, 1996; Gilbert, 1989; Carroll, 1985; Ho, 1976; Erikson, 1965; Leach, 1954). Shame is the conceptual/social dimension of embarrassment but clearly has its roots in the biological generation of embarrassment and even in guilt. Embarrassment is usually transitory, while shame is more of a social, if not theologycal, issue. But if circumstances persist in which the two are common, they feedback on the development of guilt, further directing and enhancing the taboo and boundary building processes. As with guilt, however, they can become extensive to the point where they begin to generate behavioural depression, that is, a withdrawal from social encounters motivated by feelings of despair.

Depression

Bio-electrically, depression is linked to particular neural inhibitions and a slow down and general loss of cognitive and behavioural efficiency (Mahendra, 1987; Gilbert, 1984; Gershon and Rieder, 1992; Greenfield, 1995), and it could be speculated that depression evolved from earlier systems which provided for sleep and/or hibernation. That is, it is possible that depression in humans is an evolutionary descendant of mechanisms in other species which conserve energy and take individuals out of action during hard times and/or during conditions of probable danger. Certainly the transmitter serotonin which both induces sleep and makes it hard for us to hold on to specific images (that is, maintain conscious coherence) is involved in depression (Greenfield, 1995).

Depression might logically seem to be at the opposite end of a continuum from mania/excitement. However, as a clinical or behavioural concept (that is, as a mood rather than a neurological firing rate concept) this dichotomy can be misleading. This is because both mania and behavioural depression are mood conditions in which insomnia, poor appetite, weight loss, increased libido (in some), mental hyperactivity, irritability, aggression, anxiety and agitation are often common (Mahendra, 1987). During most states of clinical depression, individuals experience high levels of mental and visceral stress-related activities which seem, in fact, to block, confuse and generally impair tendencies for total neural slow down - such as might be found in deep sleep and/or hibernation, for example. It might be that, given human awareness and the problems often encountered in reproductive and social competitions, selection resulted in mechanisms which kept individuals mentally alert without activating physical responses. This might explain, for example, why the sleep cycle is affected by depression. Depression in this sense would be a sort of awake mental play in which the mind raced hither and yon but the body remained still.

Or perhaps we should come at the question from the other end. It is possible that much of the excited behaviour common among most primates, and certainly among our nearest relatives, chimpanzees, during times of stress was repressed as part of the general repression of largely more deterministic instincts, tastes, emotions and desires. Selection, in this scenario, piled on repressive mechanisms which retarded physical behaviour without completely doing away with alertness. Often, however, the result was/is conflicting tendencies: on the one hand to act quickly, decisively, angrily, viciously, and, on the other, not to act at all, to hide, to abscond, to freeze, to avoid, to sleep, to get away.

Whatever the evolutionary scenario, depression deeply involves the limbic (emotional-cognitive) system of the brain. And it does seem that its bio-electrics involve a general firing confusion at the level of neural interconnections (above references). This happens, perhaps, in the manner of cross-firing, short circuitry, or some such, that clearly reduces short term learning efficiency (the concentration necessary for the creating and charging of knowledge units) as well as motivation for physical activity. As a result, depression acts not only to motivate individuals to withdraw from specific situations and encounters but also, perhaps, to re-think self in relationship to them. Depression hurts, it often hurts badly. Images of self are clouded with a dark gloom, a feeling that there is no purpose to life, that it is not worth living.

Depression in this sense can be a major long term motivator for a degree of emotional-cognitive reconstruction in order to get rid of the hurt, and, as a result, for reconstructing taboos, social roles, boundaries and rules. Images of previous life circumstances are recalled, however unconsciously, from deep, perhaps subconscious, memories, to be discarded, or at least recharged with altered emotional significance, owing to, not despite, the pain of depression. The relationship between depression and such 'creativity', however, is far from clear. Indeed the case is not proven. But at the same time there is no evidence that, short of chronic states, depression hinders an individual's reproductive, or indeed social, success (Mahendra, 1987; Gershon and Rieder, 1992). In fact, mild symptoms of depression can act as "care-eliciting" signals and more severe symptoms as a "cry for help", both of which are often difficult for significant others to ignore (Barash, 1981, p. 217).

Depression is very often set in motion by "a threatened or actual loss" of a significant other or of social respect (Brown and Harris, 1978; Mitchell, 1975). Feelings of loss may be exceedingly subtle, ill-defined or barely conscious " . ..that of feeling devalued and losing a sense of being appreciated, or being loved as a person" (Mitchell, 1975, p., 10). To some extent fears of loss are related to individuals' perceptions of the safety or dangers in their social universe; for example, depression is highly related to infidelities by partners, to not having trustworthy friends, to being in powerless circumstances and so on (Belle, 1982; Seligman, 1975; Mitchell, 1975). Many of the characteristics of jealousy - both bio-electrically and behaviourally - are also found in depression.

Depression may regain the attention of lost significant others, but alternatively, as noted, it can be a way of "emotional freezing when the danger and pain of loss or separation are too much to tolerate" (Mitchell, 1975, p., 27), or be a means of drawing back, for reconsidering (Gilbert, 1984; Taylor and Brown, 1988), for moving to social relationships where jealousy, the fear of envy, guilt, and other painful experiences are less likely. Depression can act as a basis for deferring gratification, in that in order to escape it fantasies of fantastic satisfactions in the future are generated; it may even become a badge of courage. Nevertheless, much of human behaviour and many human generated taboos, social rules and boundaries (often unspoken ones) are predicated in the hope of avoiding circumstances in which individuals would have to end up feeling depressed (as well as for avoiding envy, jealousy and hate). As such depression too is a major social emotion.

There are, then, a number of emotional-cognitive processes which have meaning as motivational agents in human behaviour, and so, along with their seeming universality, can be considered prime candidates for being *species typical* emotional-cognitive motivators. It would be all too easy at this point, however, to fall into the teleological trap of arguing that such species typical emotional-cognitive processes evolved to satisfy basic human 'needs' or are adaptations. However, it is not so easy and, anyway, this view would be in contradiction to the attempt to formulate a non-teleological concept of human motivation as is being undertaken here. That is, it would contradict what we know about how the evolutionary process works as a process of random change building upon what has gone before.

It is true that human emotional-cognitive processes have evolved to be in some cases extremely complex, operating as more than semi-automatic forces reacting to specific stimuli and/or mental images. They are often 'organized' - by and in conjunction with a number of more advanced cognitive processes - into relatively complex (but also potentially uncontrollable) patterns. The more complex of certain of these can be called species typical desires and fears. But the notion of desires and fears involves more than greater complexity, it, in fact, takes us to another conceptual dimension in our quest to understand motivators and repressors of human behaviour. This dimension does not mean, however, that desires and fears are a product of design or even of an evolutionary drive to greater and greater adaptation. Indeed, almost the opposite. And, it is as such that desires and fears clearly demonstrate the possibility of discovering a non-teleological model of human nature. Let us look at desires and fears in turn.

THE EVOLUTION OF HUMAN DESIRES

In a teleologically deterministic sense, we say, for example, an animal eats because it *needs* food to live. But this is wrong. An animal eats because it is *hungry*. Hunger is an internal agitation activated by a prolonged lack of food. However, in the case of humans, as Freud (1976, Chap. VII, 1971, 1962, 1957, 1984) argued, a regular satisfaction (rectifying internal tension) of a 'need', such as hunger, leads to 'positive' memory traces (images) of that fulfilment. This includes 'somatic memories' of the behaviours and objects necessary for fulfilment. As a result, Freud considered that humans can be said to have 'needs' which seek fulfilment and which 'teach' us to want what would satisfy them.

However, in considering psychological and social 'needs' he described them, variously, as: instincts, instinctive needs, drives, 'somatic-needs', 'infantile wishes' phantasies, and sometimes desires – in the form of "cathexes of energy" which sought discharge (fulfilment), but which were buried deep in the *unconscious*. In other words, although they were extremely important as motivators of behaviour, they were most often not subject to conscious control or even understanding. Moreover, they were subject to a number of "vicissitudes" and could result in anxiety and neurosis just as much as they could in happy daydreaming or 'positive' action. They had, in large part, a phylogenetically derived life of their own. Human behaviour, thus, had to be explained in terms of the working of a number of *unconscious* motivators. This was not going to be easy because not only were such motivators not readily perceptible to consciousness, they often resisted conscious understanding. Nevertheless, in attempting to build on Freud's considerable insights, we can alter his model somewhat in order to use its basic ideas. The suggestion is that we can identify a 'desire process' which in fact, goes some way to bringing unconscious motivators to the surface.

The Desire Process

In terms of the perspective being developed here, the charging process during the creation of certain knowledge units generates considerable excitement and heightened sensitivity while also dissipating tension/anxiety. As a result, during the creation of certain knowledge units (that is, during dreaming, daydreaming, thinking and fantasizing) certain *images* of behaviours, objects and persons, or combinations of these, become greatly sought after. This is the basis of the desire process. Desires are a more sophisticated version of addictions. Desires are generally more cognitively ordered and usually contain more 'ingredients' than addictions. More and varied daydreaming and thinking go into desires and they are somewhat more under conscious control than addictions. Most importantly, however, is that in desires it is often the *images* which need to be created and maintained rather than particular stimuli. Still, the difference is more a matter of degree, or intricacy, than of kind; the addiction process is clearly involved in the creation of desires.

Desires have an evolutionary basis in so far as they represent the evolution of motivational mechanisms which are more flexible and sensitive to environmental changes than instincts and addictions but at the same time considerably more consistent and very much faster than trial and error learning. Despite a degree of flexibility, human desires are not random; as with addictions, only certain objects and/or behaviours set the desire process in motion. This is because objects of desire are also to a significant extent specified by pre-knowledge units and unit potentials (established by natural and sexual selection). However, reactions to stimulating objects, behaviours and individuals vary considerably more as a result of the desire process itself than is the case with drives, instincts and addictions.

A major significance of desires is that they do not come from what modern human rationality or 'objective', scientific studies of what humans need, or should need, might suggest. The evolutionary approach to desire makes the concept non-teleological, and it explains why quite a bit of human behaviour might appear maladaptive to modern rationalists. But, above all, for theoretical purposes, it is very feasible to argue that it is the desire process rather than rationally discovered 'needs' which largely determines human behaviour. That is, wanting to maintain and to fulfil desires is most often a much stronger motivator of behaviour than striving to satisfy objectively identified human needs. Indeed, the argument here is that definitions of human needs largely come from the desire process itself.

So, for example, we 'need' oxygen considerably more than we 'need' to look nice or to be loved. But we desire to look nice and to be loved. And a great deal of our behaviour is motivated by the fact that we desire to look nice and to be loved and almost none on the basis that we need oxygen (Symons, 1979). We did not evolve a desire for oxygen because it was abundant and only a limited amount can be consumed; it was not an evolutionary problem. On the other hand, we evolved a desire to look nice and to be loved because both conditions are not (and, we can presume, were not), for the most part, automatic. They require significant effort to achieve and can easily be lost. They are relative, so an upper limit - or even a 'sufficient' amount - cannot easily be known to have been achieved. But, generally speaking, the more effort made to look nice and to be loved the greater is (and was) relative reproductive success.

However, as with addictions, this process can lead to an evolutionary trap. Let us use the analogy of the evolution of an addiction to sugar. We can relatively easily become addicted to sugar bearing foods because sugar tastes good, not because we understand that most such foods give us energy. We are motivated to eat (desire) certain foods because they give us consider-able bio-electric reward separate from knowledge of their food value. This bio-electric reward system evolved in a context in which general primate metabolic processes were able to extract and process considerable energy from sugar bearing fruits but fruit, while available, was relatively scarce (Barash, 1982). Vinegar, on the other hand, does not taste good. Vinegar has very little food value for humans; in fact it tastes much like a number of substances which are not yet ready to eat and/or are full of poisonous dang-ers for humans. We can easily develop an addiction to sugar but not one to pure vinegar (at least on its own), regardless of how much someone might try to teach us that vinegar is good for us, because we evolved a *distaste* for vinegar-like substances.

So far so good; we are motivated to spend a considerable amount of effort trying to find and to process foods based on sugar (carbohydrates) and relatively none at all on foods which taste like vinegar. However, a strong addiction to or desire for sugar can result in gorging oneself with it long after other, less harmful sources of energy become available. And a distaste for vinegar can discourage us from taking certain medicines which are beneficial. This is the case even when the analytical parts of our minds clearly tell us that the new source of energy and medicines might be better for us than refined sugar. This process represents a common evolutionary 'problem'. Selection most often favours bio-electric processes which provide motivations for the most rapid and accurate behaviours possible for survival and reproductive success in given circumstances. Selection thus favours instincts, drives, tastes, distastes and, in humans, a strong tendency for fantasies, addictions, desires and fears rather than a compulsion to more detached analytical understanding or trial and error learning (Symons, 1979). As we have seen, this is especially the case if the objects and behaviours 'needed' are scarce and competition for them is intense. This becomes an evolutionary trap, though, when animals are rendered slaves to their instincts, drives, addictions or desires (or fears) and environmental conditions significantly change. This is especially so if the change is relatively rapid.

In the human desire process there can be additional complications. The knowledge unit creating process (that is, dreaming, daydreaming, thinking and fantasizing) can become as, if not more, enjoyable than actually fulfiling the objective conditions of the images generated. Desire creation, thus, becomes very compelling (addictive) in itself. This can be the case however appropriate or not particular objects or behaviours of those images might be in given conditions. Daydreaming about looking like a princess or prince, for example, can become more fulfiling than actually making efforts to look nice. This, of course, can have quite 'negative' reproductive consequences. Also, because an upper limit, or even sufficient amount, of desire contents can not easily be known to have been achieved, there is a tendency in the desire process for individuals to continuously strive for more and more of a given object or behaviour. There is, thus, a tendency to never be satisfied, consciously or unconsciously, with what has been achieved. While in some circumstances this might enhance reproductive success, in others it can lead to obsessive behaviour which is out of step with objective conditions.

Despite the potential evolutionary trap of escalating desires, however, motivating genes usually serve a variety of phenotypical functions and so selection does not easily eliminate genes just because parts of their messages become counter-productive. This is partially because mutations which would eliminate such genes would also eliminate the other functions which these genes provide. Therefore, selection has favoured other genetic mechanisms for overcoming the dangers of extremes: the evolution of repressing, modulating and regulating genes/and or repressing, modulating and regulating functions of existing genes (Ayala, 1978; Fitzgerald, 1978; Harris, 1980; Changeux, 1980; Ridley, 1999). As a result, however, genes for potentially disadvantageous extremes will remain in a gene pool. This is especially the case where 1) mixed dominance exists and the heterozygous condition has an advantage over either homozygous condition 2) several genes are involved in a particular outcome so that most combinations are not disadvantageous - and also there is simply not one gene to eliminate and/or 3) there is an evolutionary advantage in genes providing for a range of possible phenotypical and behavioural outcomes - specific genes on the range becoming active in response to varying stimuli from specific environmental conditions, but in certain combinations (concentrations) being potentially lethal.

The extent of the significance of this for our understanding of basic human social and political behaviour can be considerable. Here the crucial point is that the human capacity for *escalating* desires (and fears) represents an additional dimension to instincts, tastes or drives in terms of the evolutionary trap problem. On the positive side, desire (and fear) processes motivate on-going evaluations of possibilities within physical, social and political environments through their tendency to be continuously 'upgrading'. In other words, it can be argued that the human capacity for potentially escalating desires and fears was reproductively valuable for an increasingly self-aware creature for whom learning through dreaming, daydreaming, fantasizing and analytical thinking was evolving. But, as has been noted, too much dreaming, daydreaming, fantasizing and thinking (such as continuous introspection) could have led to inaction at best, and reproductively disadvantageous behaviour at worst. Thus, we can theorize, at the same time as the evolution of self-awareness was taking place, the potential survival threatening and reproductively disadvantageous escalations of desires and fears accompanying this came under a degree of emotional-cognitive control.

The maintenance of alertness through something as simple as the activation of excitement or tension/anxiety knowledge units as a check on excessive heightened sensitivity (during dreaming, daydreaming, fantasizing or deep thinking, for example) might be a major means of doing this. Something along the lines of, but perhaps somewhat less dramatic than, an orgasmic discharge may be at play here. When lucid dreamers, for example, let too much passion enter into their dreams they tend to waken (Garfield, 1976; Green and McCreery, 1994); the same is generally true of normal dreaming (Hobson, 1990) where conscious control does not exist. In other words, at some point it becomes very difficult to remain attracted - or attempt to increase attachment/addiction - to specific objects and behaviours of dreams, daydreams and fantasies because the knowledge units containing their images begin to generate bio-electric distractions (excitement, anxiety/tension, depression) when excessively stimulated. This leads to a diversion away from the contents of the knowledge units, if not to making such contents seem dangerous.

More significantly (perhaps), during the evolution of the desire process, emotional-cognitive processes such as envy, jealousy and guilt were also evolving. Besides other things, these can act as repressors of potentially run-away desires and fantasies through the pain they can generate when certain fantasies become too strong; for instance, when escalating fantasies about committing adultery generate strong feelings of guilt. However, excessive *repressors* of this type can negate the advantages of daydreaming, fantasizing and thinking. So we would expect that an evolutionary conflict between the two tendencies would have gone on for some considerable time (and probably still does). The result has been the evolution of a dynamic balance between a variety of emotional-cognitive processes, desires, fears and analytical abilities. It is our species typical balance that we are talking about when we speak of human nature.

And, as a result of the precariousness of the balance, human existence is a constant struggle – personal and social; human social patterns reflect this. Dreams, play, daydreams and fantasies result in the construction of images of several desirable reproductive and survival scenarios, given a variety of changing opportunities. Dreams, play, daydreams, fantasies (and gossip) provide innumerable, continuously updated evaluations of those one meets in significant contexts and of our desires (and fears) concerning them. Emotional-cognitive processes such as love, jealousy, envy, guilt and feelings of depression hold reign on desire development, and on self's behaviour. They focus and guide fantasies, desires and behaviour on the basis of phylogenetic 'experience'. When the 'right' combination of circumstances and persons appears, however, motivation to action can be quick and decisive, not hindered by the constraints of trial-and-error learning or of philosophical contemplation and prevarication. Lust and love, for example, often act unhesitatingly, abandoning reason, when desire is strong enough and repressing emotional-cognitive processes are minimal.

Desires and Human Motivations

As noted, an important consequence of the conceptualization of desires suggested here is that it helps explain our observations and experiences that the actual acquisition of desired objects and the achievement of behaviours of fantasies and full-blown desires often fall flat. At best they only partially satisfy our desires (the emotional-cognitive rewards, remember, are often in the processes of desire creation itself). We wonder what the fuss and anxiety were all about. 'Is this all there is to it? Was it worth all that worry and trouble?', we ask ourselves. Humans, thus, not only continuously re-evaluate their environments but also are motivated to adjust their behaviour. When achieving an object or completing an activity 'falls flat', for example, we might reject the value of the objects and behaviours, including our fantasies of them. On the other hand, we might idealize them as *fantasized classes* of objects and behaviours, feeling that our particular objects and behaviours of achievement are of inferior quality, not typical of such objects and activities in general. In the first case we move on to consider other objects and behaviours. In the second case more and more, or improved versions of such objects and behaviour are deemed necessary for fulfilment. In both cases we are motivated to change either our behaviour or our environment, if not both.

Satisfaction, however, can remain elusively one step ahead of fulfilment. Individuals, thus, very easily learn to want, and become obsessional about (strongly addicted to), objects and behaviours which have existence, or at least the ability to satisfy, only in their fantasies. The alleged power for 'good' of such objects can easily become impervious to empirical refutation.

So more and more of the objects and behaviours desired, or new improved objects and behaviours, become 'needed' in order to maintain even a degree of bio-electric tranquillity or to satisfy, however temporarily and fleetingly, a fantasy addiction. This makes humans both accumulative and creative, social behaviours of great consequence in understanding the evolution of human social life.

But it does not have to be so dramatic, a capacity for habituation to certain desires and fantasies, even though the contents of the desires and fantasies may not have been achieved, is also a major motivation to look for new objects and patterns of behaviour. A sudden increase in neural inhibition and an excess of brain opiates, we can speculate, result in a loss of concentration. The mind elaborates the desire and/or jumps to others. In other words, at some point it becomes *boring* to daydream and have fantasies about achieving certain things. Even when some desires are temporarily fulfiled, considerable contact with the objects of them and/or an exercise of their related behaviours can make the objects (and indeed the desires about them) boring. Boredom, thus, can result in change, new foci for the mind, new developments in social behaviour.

In humans, then, boredom can be a major repressor of excessive desire which can have significant consequences for social change. But it must be kept in mind that boredom does not act independently of the other emotional-cognitive processes discussed earlier. All the mechanisms concerned would probably have evolved more or less together; certainly they work together in modern human nature. Boredom tends to reduce alertness, for example, and in the long run other mechanisms which checked the growth of desire while keeping the creature alert and able to divert attention to the most favourable alternatives would have been even more favoured by selection. A desire for a love object, for instance, may become boring after a certain time, but it is more likely that feelings of danger from potential rejection and humiliation may generate sufficient bio-electric tension/anxiety to repress love fantasies. Or, feelings of *guilt* for the seeming power of the desire, or for experiencing lust jealousy, anger and hate (feelings that love fantasies often generate), may divert the desire process in other directions. Another possibility is that boredom generated fantasies are shattered by the fantasy being unexpectedly fulfiled - sex with the neighbour's husband, for example - but the result generating a sufficient level of guilt (let alone fear of being found out) so that all fantasies which have an element of adultery attached become painful, tending to hinder such fantasies.

As a summary example of much of what has been considered in our discussion of fantasies, addictions and desires (and fears) with regard to human motivations let us look at the case of male masturbation and the generation of male sexual desires. It can be argued that the evolution of a capacity for a very easy to acquire masturbation addiction, and for masturbation fantasies, helped motivate mother dependent human males to leave their mothers. This would be because masturbation fantasies divert male infant intimacy/sexual desires away from mothers towards non-incestuous matings (Friday, 1980) - there being, presumably, a block on masturbation fantasies which include sex with mothers. However, an evolutionary result could have been that a considerable amount of time was spent in masturbation *con* fantasy and none in more direct reproductive pursuits.

Therefore, we would expect that mechanisms which repressed motivation for masturbation and pleasurable masturbation fantasies, in favour of sexually reproductive behaviour, would have evolved. Masturbation repressors probably include: 1) a degree of boredom after a certain age (but probably not); 2) a fear of pollution born of disgust reactions to being covered with the sticky, obtrusive, clothing discolouring, slimy, potentially unhideable nature of the result; 3) or, conversely, a fear of a loss of 'vital' bodily fluids/powers/juices; enabling 4) an easily learned fear that masturbation will cause insanity (or at least feeble-mindedness or blindness); 5) an easily developed sense of guilt for finding it so difficult to stop *despite the clear dangers*; this sense of guilt is usually intensified by 6) conflicts between masturbation fantasies in which sex with sisters, aunts, female teachers, friends' mothers and Mrs. next door neighbour abound and a fear of these seemingly uncontrollable 'perversions' slipping out or being discovered. Based on these repressors humans 7) often generate hygienic theories which postulate causal relationships between masturbation and general poor health, weakness and/or infertility; 8) religious teachings which clearly portray masturbation as a defiling, sinful activity; in a context in which 9) masculinity and sexual conquest desires soon equate masturbators with 'can't do men' ("wankers"/"jerks") – 'can't find a woman dodos'.

As repressors of masturbation, these mechanisms would, of course, be supported by evolving desires for copulation. But in real life, we must remember that only some of these are going to work and at best will only be more or less effective. Or, in some circumstances they may work only too well and all fantasies concerning sexual activities will become greatly restricted, if not blocked altogether. The mechanisms for developing desires and for their moderation are far from having achieved a perfect balance. This may be bad news for those who want to be able to achieve supremely adapted, healthy selves. But from an evolutionary point of view it makes complete sense. In general, most environmental conditions, physical and social, select for a degree of flexibility of response in relationship to a range of changes within those conditions. Evolution, thus, results in a range of possible individual reactions. In humans, this means that in any given individual mixtures of mania, obsession, addiction, habituation, boredom, anxiety, tension, depression and specific desires (and fears) will vary from circumstance to circumstance. Variation will depend, at the very least, on specific environmental conditions, pre-knowledge and unique individual experiences.

In other words, degrees and types of individual motivation will depend on specific external stimuli, the existence of specific pre-knowledge units, knowledge unit potentials and knowledge units created through experience, as well as on the degree and type of bio-electric charging in these. All of this, in turn, will have been affected by past experiences and past dreaming, daydreaming, fantasies, thoughts, desires (and fears) of a given individual. This means that within a given population there will be a range of process represented in a number of different individuals. Generally, however, the *ranges* within individuals will fall somewhere within the arithmetic mean of the range for a population. This is because the middle of the population continuum will have produced the highest level of reproductive success in most circumstances. Thus, to an extent the sharing of this range could be said to define the modern species of humans.

In summary, then, during human evolution selection would have favoured the intensification of some existing primate emotional-cognitive processes and desires (and fears). It would also have selected for the development of some new ones. In other words, during human evolution new genes, or new parts of existing genes, determined subtle changes in bio-electric mechanisms and processes in our brains which made it bio-electrically rewarding to dream, daydream and think about, and to be attracted to and to play with, a number of new behaviours and objects.

On the other hand, some primate drives, emotional-cognitive processes and desires would have been counter-productive from the start and these would have been bio-electrically repressed during the evolution of the human emotional-cognitive system. Moreover, in this highly conscious, daydreaming species, some desires in the process of intensification would have 'gotten out of hand'. At some point selection would have favoured controlling and repressing mechanisms for such counter-productive desires. As noted, there are a number of neural processes and species typical emotional-cognitive processes which do just that. Many of these result in the generalized controlling mechanisms we refer to as human fears.

THE EVOLUTION OF HUMAN FEARS

We have already seen that stimuli derived from certain objects and behaviours lead to bio-electrically felt feelings of danger (possibly through conflicts between excitation and inhibition at the neural level). When the above danger inducing stimuli - from external and internal sources - are dreamed, daydreamed and thought about, they become more than activators of bio-electric precursors to flight, freeze or panic reactions. They are on the way to becoming human fears, a process completed when these feelings are bio-electrically charged and, to some extent at least, cognitively ordered. During the process, some stimuli generate 'negatively' charged knowledge units and some activate pre-existing 'negatively' charged knowledge units, that is, units with branching to tension/anxiety/depression mood conditions. While doing so, we can speculate, frantic-feedbacks are sometimes generated with conflicting tendencies toward, at the same time, excitement, depresssion and mania.

The mental hyperactivity and confusion caused by the generation and activation of fear knowledge units are viscerally felt as physical tension, pain, illness, embarrassment, shame and mental confusion. These are usually expressed as frustration, disgust, terror, visible anger, aggression, despair and clinical depression. The activation of fears motivates individuals to avoid, to worry about, to plot against, to impose power upon and to become mentally and physically immobilized by stimuli from certain objects, behaviours, fantasies and thoughts. During the splitting, classifying, stereotyping and stigmatizing common to more advanced cognition (see Chapter Six), certain types of behaviour, thus, come to be designated either safe or dangerous. Safe knowledge units are usually linked with conditions felt to be necessary for desire fulfilment generally. Danger units are linked with images of conditions in which desires are not likely to be met, emotions are uncontrolled, desires are out of hand and/or enemies (physical, sexual, social and political) are likely to abound. So danger seems to have a life of its own. The world comes to be seen in terms of good and evil and/or of purity and pollution; evil (persons, objects and events) and pollution (persons, objects and events) often seems to have a power greater than ordinary humans.

In other words, when daydreamed and reflected upon, causes of fears (that is, what seem to individuals to be the sources of dangers) become the basis for human notions of evil, wrong and impurity. Moreover, as with desires, fear reactions can become much more than objective evaluations of personal conditions, behaviours or other individuals. They generate stereotypes which include images of hypothetically dangerous objects, people and behaviour, of hypothetically impure objects, people and behaviour. Fears spark fantasies of inhuman villains and enemies, of devils and witches. The above classifications of acceptable and unacceptable objects, people and behaviours, and their links to charging processes, are often stored in long-tern memory.

A considerable amount of bio-electric pain, thus, can be experienced when daydreaming and thinking about dangerous objects, individuals and behaviours. As a result, fears, like desires, can become greater as motivators than the effects certain objects, people and behaviours can, from an outsider's point of view, cause. This in an extreme case often leads to guilt and depression, for example. And, as with excessive desires, an excess of fear can be reproductively counter-productive, can become an evolutionary trap. Therefore, it is not surprising that selection provided mechanisms for reducing excesses in both. As with desires, certain fears can be habituated and even become boring. Left hemispheric analytical processes often reduce the effects of more viscerally developed fears. This is done, for example, via the power to control and tame though classifying and naming. It is also done via analytically defining excess emotional expressions and certain desires as being dangerous/evil, and thus to be collectively guarded against.

Thus, analytical-cognitive processes can both exaggerate and dampen the escalation of human fears. All of this motivates the control of excessive behaviour in self and in others. Socially, fears motivate us to try to control, organize, predict (Rachman, 1978) and prevent the occurrence of behaviours and activities which activate our danger mechanisms. The human danger system, and its product, human fears, has evolved in order to guard us against certain objects, feelings, motivations and behaviours, individual and social, which are likely to be reproductively disadvantageous and/or dangerous to our survival (a major task in an evolutionary social science will be to identify some of the possible objects and behaviours which trigger danger reactions).

Since this often requires both the control of certain emotions and desires in self and in others, it implies self-control and the control of others. Much of this happens with the aid of the bio-electrical mechanisms of some of our species typical emotional-cognitive processes (for example, love, guilt, envy, control, empathy) which also underpin, in part at least, the human capacity for deferred gratification. In general, human fears act as major motivating mechanisms for the psychological and social processes underlying the establishment of taboos and social boundaries (see Chapter Seven). Human behaviour is thus a product not only of species typical pre-knowledge, emotional-cognitive processes and desires but also of fears (along with species typical patterns of advanced cognition). These all together, along with some more advanced cognitive processes (see next chapter), make up the natural kind of human nature.

SUMMARY AND CONCLUSIONS

In this and the previous chapter I have sought to outline a possible model of the workings of the human emotional-cognitive system, or systems. The advantage of this model, I suggest, is that it is clearly linkable to the workings of neural bio-electrics (which in turn are linkable to genes) on the one hand, and to fuller human cognition (and these together to behaviour) on the other. Moreover, at every stage various systems are pictured in such a way that they have both a core of consistency (genes, pre-knowledge) and developmental aspects (knowledge unit potentials, images, desires and fears) which must interact with variable environments (external stimuli) in order to achieve any sort of effect.

The model, thus, attempts to integrate the obvious fact that humans learn their behaviour, with the equally obvious facts that some things are considerably easier to learn than others and that some things are much easier to retain in memory than others. The model suggests that species typical emotional-cognitive processes *charge* species typical analytical orderings, resulting in species typical memory and motivations for social actions. Through play, dreaming, daydreaming, fantasizing and thinking, the objects and behaviours which an individual encounters are analyzed in terms of human pre-knowledge, species typical emotional-cognitive processes, desires and fears. They are subjected to an 'analysis' in which they are perceived to have rewarding and boring, but also potentially dangerous, aspects. They are also subject to change and development.

Consideration so far has focused on the bio-electric charging and motivating side of the human nature equation, and as such suggests a considerable number of potential contradictions within a given individual's emotional-cognitive processes at any one time. For example, certain internal agitation or external signals could trigger unrepressed sexually oriented feelings (lust, love) while other signals trigger repression of these (for example, guilt, embarrassment, shame). All of these are possible at the same time, in a given individual, as presented in the framework of emotional-cognitive processes suggested here.

Thus, a causal analysis of human emotional-cognitive processes running from perceptions (stimuli), to bio-electrics, to behaviour, quite apart from the existence of fantasies, desires, dangerous desires and fears, suggests a multitude of processes, many of which can be operating at the same time. This makes it very difficult to conceptualize a unified, non-conflicting ego, self, species being, or whatever else we might want to call it. (In fact this probably is a very important lesson to learn about basic human nature.) Nevertheless, human emotional-cognitive processes are perhaps less conflicting and somewhat less un-systematic than the above observations might suggest, as we will see from our consideration of advanced cognition.

CHAPTER SIX
ADVANCED COGNITION

Advanced cognition (see note three, p, 234) comprises those mental activities in which objects, behaviours and individuals are identified, defined, evaluated and categorized. In the process a number of species typical binary divisions are created. Holistic, hierarchical and stereotypical evaluations of what these depict are generated. Advanced cognitive processes also attribute agency (cause) within social relations, resulting in species typical patterns of deterministic thinking, identifications - with objects and others - and teleological reasoning. Deception and self-deception are also major features of advanced cognition.

Bio-electric (emotional-cognitive) charging during advanced cognition results in propensities for specific (species typical) behaviour. These, in diverse combinations, generate strong attractions and repulsions with regard to specific objects, peoples, groups and behaviours (taking the social form of, for example, taboos, rules, social boundaries). The result is identifiable patterns - or 'natural kinds'- of human social behaviour (see next two chapters).

In this chapter I will undertake a detailed consideration of the mechanics of advanced cognition as described above. It must be noted that the use of the term 'advanced cognition' is meant to represent those aspects of the human mind which seem to have evolved later than basic emotional-cognitive processes. Advanced cognition is taken as an evolutionary elaboration of these rather than in any sense being an argument for the evolution of 'rational man', somehow above biology. Nor is it meant to imply a moral or technical superiority of cognition over basic emotional-cognitive processes.

EVALUATING

Binary Divisions

Humans have a natural ability to imagine what stands in opposition to what is observed and/or imagined. This is done, we can hypothesize, when stimuli more or less simultaneously activate pre and developing knowledge units which seem to be in conflict with or completely different from each other. These oppositions, we can further postulate, affect the generation of more specific, situation specific, opposing knowledge units. This, as with pre-knowledge and knowledge development generally, could take place during protein synthesis, divergent circuit development, glia cell formation and/ or the growth of micro-columns. Whatever the specific bio-electrics, the result of this aspect of human cognition is that representations of the envi -

ronment come to be more or less organized around a number of binary oppositions, for example: hot/cold, high/low, strong/weak, smart/ stupid, ugly/ beautiful, human/animal, mind/body, them/us, good/evil, true/ false and many more (Campbell, 1984; Leach, 1967; Levi-Strauss, 1966; Kelly, 1955; Haraway, 1989).

It must be noted that the above examples of oppositions are not necessarily the ways in which philosophers might define conceptually or linguistically pure opposites. It can be argued, for example, that the linguistic/conceptual opposite of ugly might be plain, pleasant or not ugly rather than beautiful, or that the opposite of mind is matter. This, however, is of little concern; it is playing with words. There is no reason, for example, to believe that human cognition represents an evolution of mechanisms for being able to formulate linguistic/conceptual purities which reflect a reality (rationality or higher truth) to be discovered. Nor is it likely that being able to agree on meaning will give us a greater understanding of human motivations or intentions.

Rather, it seems that humans have a natural tendency to classify the world in terms of binary opposites, and, moreover, that there are probably more or less pre-determined ways of doing so. The above oppositions represent species typical oppositions, as expressed in terms of the linguistic/ conceptual conventions of modern Western, English speaking societies. Humans who use them know what they mean. For example, mind/body, or mind/matter will both have meaning which will be understood because of the context in which it is used and because of the emotional elaboration and non-verbal communication which goes with it. In other societies, or times, different words for the above distinctions might be used, and some distinctions emphasized more that others. Perhaps in some cases even slightly different orientations to various distinctions might exist. But the ability of distinctions to convey species typical, that is, relatively consistent, basic survival and reproductive information remains the same.

These types of oppositions, as commonly understood, make up the bases around which more complex classifications are developed. Thus, human cognitive processing can refine these simple binary oppositions with further oppositions (and this can be repeated a number of times). So, for example, the notion of hot contained in the simple binary opposition hot/ cold could be refined by, for example, opposing the hot end between dry hot and muggy hot, and further dry hot between desert hot and room hot, and so on, continually specifying the notion of hot (see Fig. 6.1)

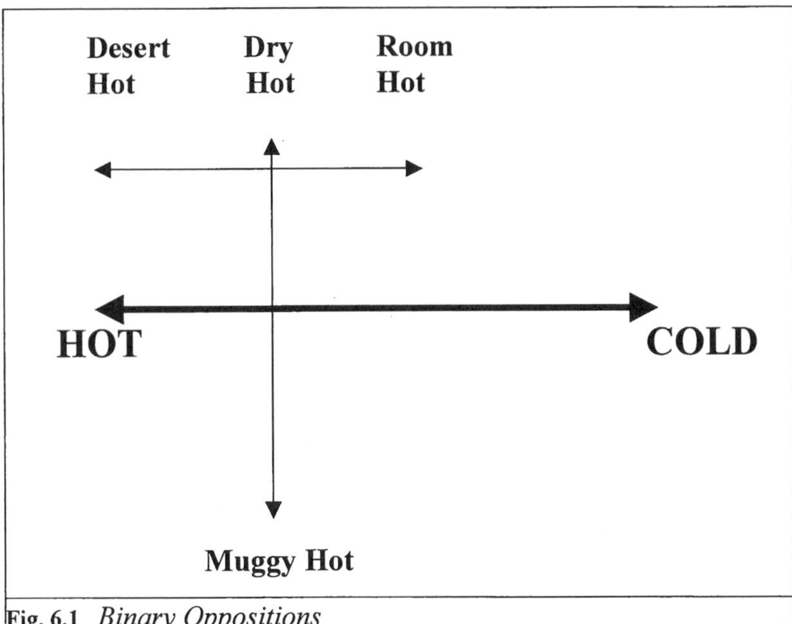

Fig. 6.1 *Binary Oppositions*

This refining of simple oppositions gives us a certain amount of useful, long term, human-relevant information which can be used to guide our behaviour; for example, it can be used to 'tell us' that, 'we will 'need' plenty of water' and 'don't stay too long in this place', or to 'turn down the heat'. This contribution, however, is greatly enhanced when holistic evaluations based on the inter-relationship of several additional environmentally relevant, oppositions are developed.

Holistic Evaluations

The right hemisphere of our brain seems to specialize in this function (Watzlawick, 1978; Buck, 1984). So, elaborating upon my example of temperature, the opposition between hot and cold (after each end of this continuum has been specified through further oppositions, e.g., between muggy hot and dry hot) can be intersected with other continua, such as between fog/clear, glaring/dull, rain/dry and windy/calm, for example (each end of these continua, of course, may have been specified). This gives us a conglomerate picture which represents a notion of weather and/or climate (see Fig. 6 .2).

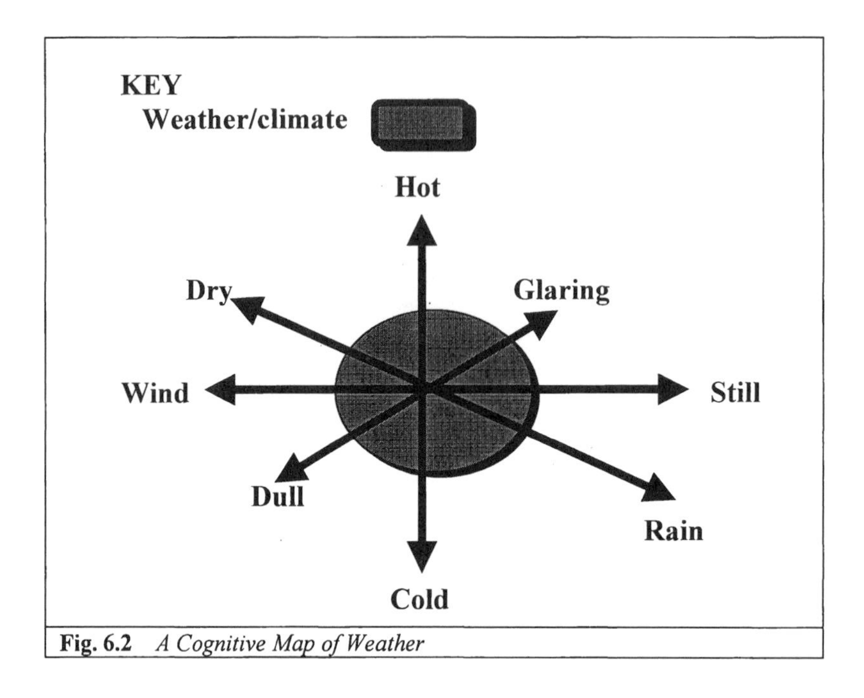

Fig. 6.2 *A Cognitive Map of Weather*

Such holistic evaluating, however, introduces another dimension of advanced cognition; one that works in almost the opposite direction to the development of simple binary oppositions. That is, one of the conflicts human nature has to contend with is that besides easily creating simple binary oppositions humans can also feel that there is a great deal of ambiguity and confusion in the real world. Often what are seeming oppositions become fused and ambiguous. There is, therefore, also a human tendency to conceptulize oppositions as being gradations. This is, one might argue, an advanced cognitive compromise between a tendency to see stark oppositions and a recognition that the world being interacted with is decidedly less clear.

Although generated from a perceptual cognitive conflict, a conceptulization of gradation, nevertheless, provides for a flexibility of evaluation of what we perceive. It gives us an ability to observe and evaluate ranges and complexities of circumstances and conditions. Each continuum in Fig. 6.2, for example, can be graded and then slid backward or forward in order to designate weather according to different environmental conditions. In some circumstances the middle point between dry and rain might be linked with the middle of other continua, but in other conditions an intersecting point much nearer the rain end would better represent local perceptions of typical weather, and so on (see Fig. 6.3).

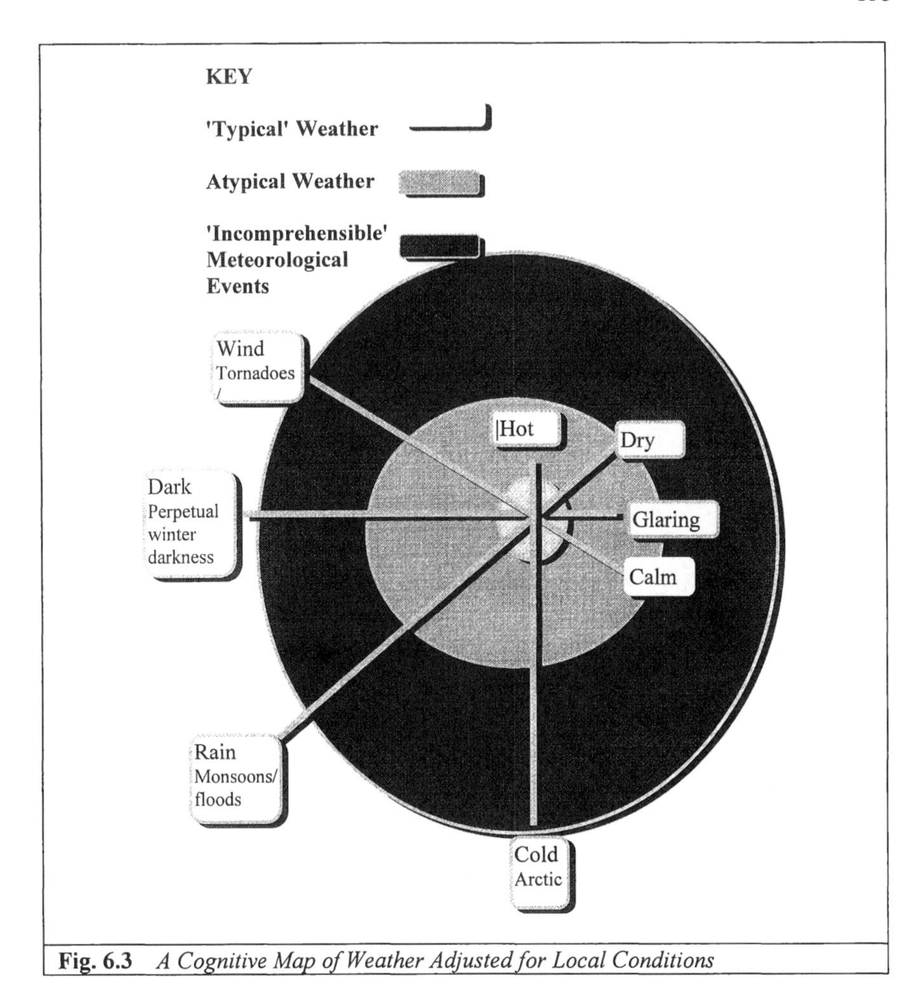

Fig. 6.3 *A Cognitive Map of Weather Adjusted for Local Conditions*

At this level of evaluation we tend to know what kind of coat might be best, whether to wear sun glasses, galoshes, snow shoes, a raincoat or a wind-breaker. This level also begins to tell us if it is a safe or dangerous environment, time to plant the crops or to hunt the reindeer. Significantly, all of this can be communicated in abstract terms so that an individual does not need to experience extremely desert hot weather, for example, to learn from others that it is unpleasant, that most crops do not grow in it, that reindeer do not live in it and, in fact, that it can be dangerous.

To some extent, then, specific conglomerations of intersecting points develop with human learning experiences; nevertheless the importance of the existence of opposing pre-knowledge units, or at least unit potentials, used in creating holistic weather images is strongly suggested by what seems to be a tendency to universal human weather binary divisions and their gradations. This is theoretically derivable from the fact that humans - and other animals – can only tolerate a relatively very limited range of climatic conditions and as a result have evolved specialized abilities to perceive their dangerous and safe ends and, of course, to evaluate conditions within them. More direct evidence for pre-existing knowledge units is provided by the fact that human perceptions of safe weather (and environments - Orians and Heerwagen, 1992) often contain, regardless of culture, at least some of the characteristics of climatic conditions similar to the ones in which humans probably evolved ('gallery forests', woodlands, savannas, temperate to tropical Africa). Safe weather, in other words, almost always includes: relatively abundant light with clear, blue skies, in relative warmth; as opposed to overcast skies, heavy rain, wind, mud or snow.

In other words, humans feel safe weather to be that in which, at an earlier stage in their evolution they successfully secured food, shelter and mates. Sunny beaches, for example, are still more desired than areas where there is a considerable amount of rain despite the fact than strong sun causes skin cancer, turns the land into an unfruitful desert and the rain is essential to the long term maintenance of most kinds of human-friendly productive environments. Nor is familiarity of experience a good predictor of what is considered safe weather. Summer is usually considered to be less dangerous than winter even where winter is twice as long as summer.

Whatever the specifics, during holistic evaluating complicated climatic conditions are evaluated, and we gain the knowledge that they are either safe or dangerous. While this is taking place, relevant knowledge units in the brain become charged with various combinations of heightened sensitivity, excitation and/or depression. This gives our holistic images of weather emotional force. We like safe - predictable, knowable, light, sunny, warm - weather; it becomes good weather. And we dislike dangerous – unpredictable, unknowable, generally dark, windy and stormy - weather; it becomes bad weather. We have thus created a new binary opposition between good and bad weather. This reduces the number of holistically combined variables previously considered (encompassing varying degrees of light, temperature, humidity, air movement, and so on) to just two opposites - good and bad weather. 'Real', local weather possibilities nevertheless remain on a continuum, between these two opposites. We have thus arrived at a relatively limited number of stereotypical ideas of weather - arranged hierarchically. In the process human cognition has evaluated considerable information and we have a basis upon which to act, plan and generally conduct our lives (reproductive and survival interests).

Hierarchical Evaluations

With hierarchical evaluations, however, three additional aspects of this dimension of human emotional-cognitive processing come into play. These aspects are much easier to understand if we move away from our example of weather and consider the processes involved in evaluations of, for example, humans/behaviour. To begin with, humans and general human behaviour can be evaluated in a fashion similar to that described above. Humans can be friends/strangers, big/little, smart/stupid, beautiful/ugly, strong/ weak, loyal/disloyal, trustworthy/untrustworthy, truthful/lying and so forth. Each end, of course, can be further opposed, thus refining each of these notions.

A holistic conglomeration of these gives us a notion of 'typical', knowable, humans. From the holistic image of typical humans or typical behaviour, we get a concept of normal humanness which becomes the basis for a new opposition - normal/abnormal humanness. The safe end of this new binary opposition becomes 'positively' charged and the dangerous end becomes 'negatively' charged; as a result, we tend to like normal humans and dislike abnormal humans. The sliding of the continua back and forth while describing a specific individual, gives us a quick - often moral - evaluation of that individual in relation to typical humanness.

At this point we have achieved the same level of conceptual development as in creating the concept of weather in our previous example. But along the way we encounter our first additional aspect involved in hierarchical evaluations. As compared to the development of a concept of weather, the safe ends of the original oppositions which define normal humanness tend to carry with them, or are given, critically stronger 'positive' charging; at the same time the dangerous ends also receive significantly stronger 'negative' changing; a 'negative' charge for the notion of a disloyal person is of a different magnitude from a 'negative' charge concerning cold temperature, for example.

However, it is not easy to know who might be a disloyal person. So the emotional-cognitive processes involved here have, through evolution, become quite complex. In this light the second and third additional aspects involved in the processes of hierarchical evaluation enter the picture. The second aspect is that from our concept of normal humans we have both the capacity and propensity to extrapolate to a concept of an ideal and then on to a saintly, perfect human or human behaviour; in other words, we create a sort of super-goodness (often by conglomerating the safe ends of our original binary oppositions). The same process applies to the abnormal end. The extrapolation here takes us beyond abnormal to bad and then on to evil (see Fig. 6.4).

In other words, the distinction between ideal and abnormal develops into a hierarchy of humanness. On this hierarchy we can very easily dream, daydream, fantasise and theorize certain humans into heroes and heroines, or alternatively cast them in the exaggerated abnormal direction, into anti -

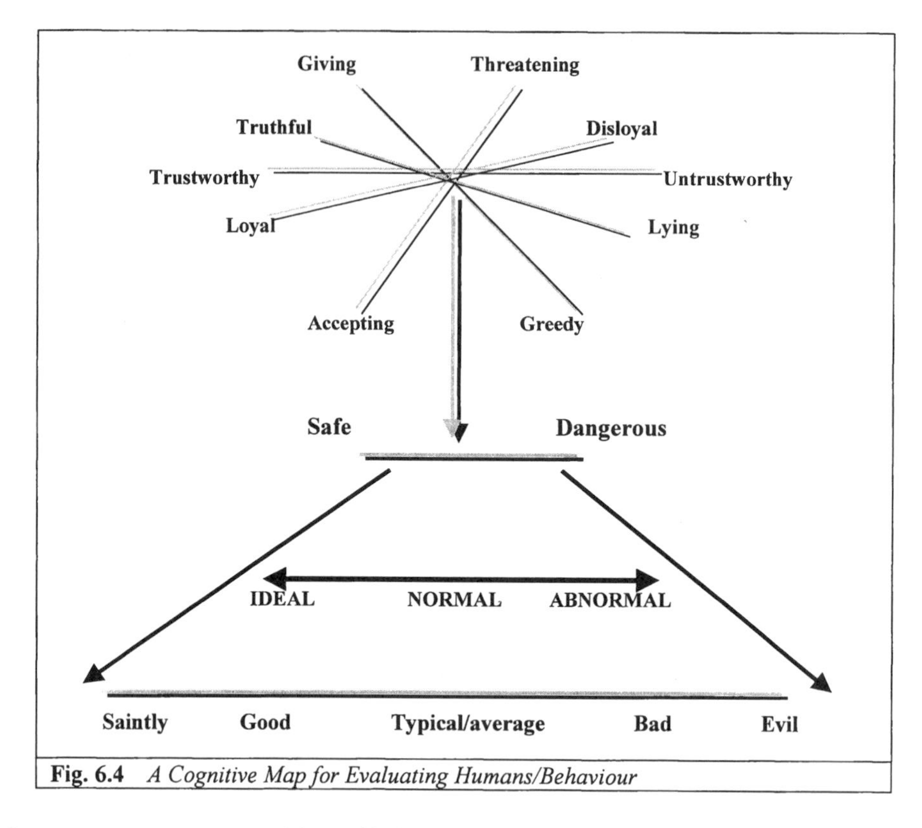

Fig. 6.4 *A Cognitive Map for Evaluating Humans/Behaviour*

humanness - represented by evil villains and witches, for example. This hierarchy also provides the basis for conceiving of perfect friends and lovers, or conversely, humans (and a display of humanness) we wish to avoid at all costs. But in all but the rarest cases neither end truly signifies our real experiences or expectations. As such, then, this process takes us beyond experiences; it provides the basis for an emotional-cognitive development of an abstract, that is, mentally embellished, conceptualization of, for example, good and evil.

When the evaluating process involves individuals we actually interact with, however, it can become increasingly complex and also increasingly ambiguous. This brings in our third additional aspect of hierarchical evaluation. As noted, the holistic image we eventually attribute to a given individual is derived from placing him or her at a specific location on a hierarchy of humanness, and in sliding various relevant secondary continua back and forth to specifically fix upon them a 'personality' – as seen by us. As we are forced, or wish, to get to know them better, further binary oppositions, such as: effusive/quiet, intrusive/withdrawn, engulfing/rejecting and doineering/submissive, for example, can also be used. This gives us a more detailed evaluation of a particular individual (as they become our friend, lover, patron, adversary). Our third additional aspect comes in when the complexities and ambiguities of these last oppositions are reflected upon. The more refined the oppositions become, the less easy it is to say, un-ambivalently, that one end is good and the other bad; which is good and which is bad, for example, between intrusive and withdrawn behaviour? Furthermore, the ambiguity of the extremes in some of these refined divisions can begin to worry humans that extremes in any form might, in fact, be dangerous.

While humans usually fear evil people in the abstract (as bullies, adulterers, mass murderers, child killers, tyrants, traitors or witches, for example), this may also pertain to the super good. In some ways we idealize smart, strong, tall and beautiful people, but in practice we are often jealous and envious of those who seem to know everything, triumph over us in physical games, steal the attention of our lovers from us - and all the while are deemed to be 'wonderful'. People who are super trustworthy, giving, loyal, truthful and accepting are also idealized at one level and feared at another. They make us look bad. So we rationalize and theorize that people with such characteristics are perhaps naive and lacking in critical judgment, and we wonder – aloud - if there is anything about them as 'individuals' other than their unrelenting tendency to 'unthinking' conformity. Do they have the courage to be even slightly, humanly, deviant, to do anything which is not one hundred percent socially approved of?; are they obsessive and secretly possessive, we wonder to our friends. We hint at the possibility that they are fundamentally hypocrites and deep-down devious.

We can even easily come to believe that the virtues they supposedly embrace are questionable. We not only never forgive the truthful friend (or acquaintance for that matter) who eschews a few white lies to tell us the brutal truth about ourselves, but we also, as a result, begin to wonder about the sanctity of truthfulness. This opens up a major aspect of human discourse, the discourse of justice. If super good and totally truthful people (moral people) can hurt us, and so seem to act very 'unfairly', the concept of morality itself begins to lose some of its spiritual significance; it no longer seems to have its necessary imperative; it is no longer incontestable. Therefore, as individuals gain family, sexual, inter-personal and political experiences, they very often subtly shift their analysis from the worth of the characteristics of persons, and the abstract value of behaviours, to the issue of 'fairness of behaviour' - as manifest in particular circumstances.

The notion of 'fairness of behaviour' has the virtue of being judgmental but nevertheless flexible. Such a concept remains open to evidence from several sides of an argument, to extenuating circumstances; it is changeable and developmental in the sense of being open to lessons from the past. It is little wonder, then, that the notion of fairness has inspired notions of justice since the beginning of written history (Cf. Ryan, 1993). And as fairness, justice is very much an issue of politics and political institutions; that is, of on-going decisions based on the sense of fairness as felt by members of a human community (Cf., Hume, 1975, Appendix III).

This suggest that a sense of injustice based on the notion of unfairness might have more fundamental emotional-cognitive roots than the concepts of evil or immorality. Children, for example, get much more upset if they think that something is 'unfair' than if they are told that it is immoral or evil. This sense of unfairness, it can be argued, derives from a fear of losing out or of being excluded. The sense of immoral or evil behaviour requires more cognitive development than this, but can be quite fickle, changeable. In fact, it may derive from the elaboration of a sense of unfairness spiked with emotional-cognitive force. Nevertheless, it does seem to be the case that there are two major processes involved in developing modern humans' notions of justice: one is a primordial fear of losing out and the second is a recognition of the ambiguity and unreality of the cognitive oppositions which describe human behaviour when elaborated into extremes such as moral/immoral, good/evil . But all of this is running ahead. Let us return to the issue of human cognitive evaluations based on a tendency to the making of binary divisions.

It is not only weather, people and behaviour which are subject to the processes discussed so far. Similarly, objects are divided into: big/little, hard/soft, ugly/beautiful, sweet/bitter, symmetrical/asymmetrical, and so on. And of course, each end of these continua can be further refined/specified. Conglomerations of intersections of various points on the resultant oppositions are used to stereotype the objects which we come into contact with. These stereotypes are then hierarchically evaluated in relation to other stereotypes and are emotionally charged and, as a result, some objects come to be highly desirable and others repulsive. (It must be remembered that the existence of species typical opposing knowledge units potentially guide this whole process, as it does with the evaluation of people.)

Again, extremes can be perceived to be dangerous. The danger of not enough food, for example, is clear but the danger of too much (gluttony) requires more elaborate cognitive development. At the same time, we often fear that certain objects are idealized to the point that at least some individuals will become obsessive about them - jewelry, elaborate decorations, guns, horses, cars, houses and expensive vacations, for example. Money, the means to and symbol of so many other things such as status and power, while idealized at one level is feared at another. We are anxious that it changes people, we do not really trust people with too much, and so on.

Activities too are divided: physical/mental, easy/difficult, possible/ impossible, fun/drudgery, safe/dangerous, fair/unfair, legal/illegal, moral/ immoral and so on. When holistic images of particular activities are drawn from intersections between refinements of these divisions, they define the parameters of what is considered to be normal for that activity. This can be formalized so that a specific conglomeration becomes the basis for delineating responsibilities and duties for designated tasks, as in the creation of professional norms, for example. When such holistic images become bio-electrically charged they are used for making judgments not only about the relative importance of various activities but also about the competence (and sometimes moral value) of those who undertake them. But again, idealized behaviours and attitudes can generate such anxieties that they are avoided. They require too much abstinence, too much concentration over extended periods of time; their practitioners are too bureaucratic, too pure, they do not allow themselves any fun, and above all, often they do not forgive failure, backsliding, or incompetence – they can be hard to get on with.

In summary so far, advanced cognition (in conjunction with knowledge unit 'hunger', addiction tendencies and the general charging - 'positive' and 'negative' - of pre-existing knowledge units – Chapters Four and Five) operates in a hierarchical fashion with regard to human characteristics and social behaviours. The pattern of this is, in a self-aware creature, the tendency to create images of ideal (survival and reproductive) behaviours to strive for and of evil behaviours to avoid at all costs. Selection, however, favoured capacities for ranges of fitness oriented behaviour 1) because of the advantage of such flexibility in changing environments and 2) because of the evolution of very complex, changing patterns of sexual competitions within a self-aware species. As a result, a check against obsessively striving for the complete ideal itself seems to have evolved. Frantic-feedbacks and general fear reactions generated in response to images of uncontrolled strength, violence, lust, anger, envy, love, jealousy and hatred, for example, succeeding against us, seem to have acted as selection forces for the evolution of a 'positive ' emotional-cognitive link (feeling of safety) with the middle of many of our species typical binary divisions - especially after these binary oppositions have been refined.

As a consequence, as an arrangement of charged knowledge units in the human mind, individuals and their behaviours are hierarchically arranged from 'the good' through typical to evil; from moral through acceptable to immoral; from God's preferences (saintly) through natural (mortal man's normal capabilities) to products of the Devil's wicked, anti-God un-natural designs. By this stage, advanced cognition is clearly involved in the desire and fear generating process; that is, it is becoming increasingly integrated with the emotional-cognitive processes which significantly motivate and control species typical human behaviour rather than in just classifying and evaluating it.

As an example we can take the development of human concepts of, and motivations regarding, sexuality, as an illustration of relatively straightforward advanced cognition becoming more complex and blending into the generation of desires and fears. Sexuality can be placed on a continuum running from reproductive to recreational. At the same time, it can also be placed on a continuum running from physical to spiritual. These, based on pre-knowledge, can be intersected in order to give a very rough idea of species typical, average sexuality. However, humans usually make a number of refinements to these binary divisions which give participants a greater amount of information concerning specific manifestations of sexuality. So sexuality is often also divided between, for example: selective/promiscuous, delightful/necessary, romantic/lustful and purifying/polluting. When these ranges are taken together and intersected and conglomerated, we get a more precise conceptualization of average/normal (species typical) sexual behaviour.

Each of the bisecting lines, however, can be moved back and forth to emphasize some or other aspect of sexuality, depending on local and/or personal conditions. In certain circumstances (that is, cultures/life experiences) the physical end and its refinements would most likely come to be emphasized, and in others the spiritual end and its refinements, and so on. Extended dreaming, daydreaming, reflecting and fantasizing will charge and fix particular combinations so that they become the sexual identities of individuals, specific sexual identities clearly depending on specific individual emotional-cognitive and advanced cognitive capacities and on individual experiences in particular environments - both social and physical. As with many other emotional-congnitively charged concerns of humans, there is a tendency to exaggerate the ends of certain of the continua so that fantasized and theorized versions become the basis for notions, at the top, of higher sexual morality, and, at the bottom, of sexual immorality, if not unspeakable wickedness such as sexual perversity (see Fig 6.5)

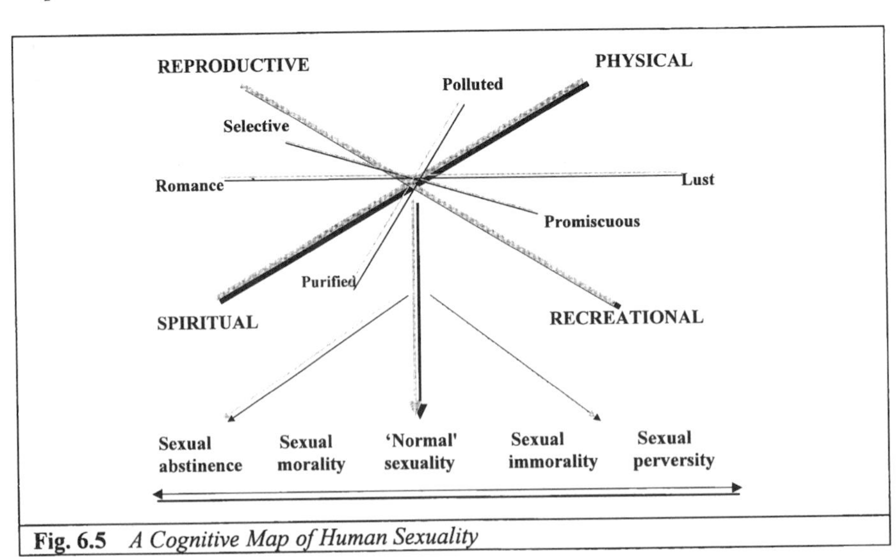

Fig. 6.5 *A Cognitive Map of Human Sexuality*

The charged idealized ends of the continua have in common that they represent lust controlling tendencies; for example, the emphasis is on sexual selectivity, romance and purity (the ideal is a sort of virginal motherhood). Their species typical opposites, promiscuity and pollution (e.g. masturbation, homosexual promiscuity and very non-selective sex generally), represent a lack of such control. Sexual immorality, when taken to extremes becomes continuous sex without discrimination (even with the very young and, indeed, with animals). This undoubtedly evolved because neither extreme gives very reliable reproductive success. Those who desire promiscuity are often avoided by the opposite sex and/or left to cope with offspring on their own; the very young and animals are not genetically compatible for reproductive purposes. So it is understandable that motivators/ propensities for sexual restraint evolved. However, sexual morality when taken to extremes often becomes counter-productive; it can becomes sexual abstinence, for example. Thus, human nature has evolved sexual propensities and anxieties in such a way that neither extreme is desired by the vast majority of humans.

In so far as normal sexuality can be considered to be a balance between extremes we can consider that there is a dynamic tension between them which maintains it. The balance, it can be argued, is caused by a predisposition to a variety of daydreams, thoughts and experiences which include often conflicting and contradictory: sexual attractions, repulsions, desire, senses of envy, feelings of love, jealousy, fears of rejection and/or engulfment and so on (see Chapters Four and Five). The evolutionary advantage of a dynamic balance was (and is) that an individual's concept of sexuality could move in either direction, and was (and is), thus, able to be generally flexible in a variety of changing environmental and newly evolving sociopolitical conditions. Nevertheless, in most circumstances, staying near the middle usually prevented (and still prevents) the reproductive disadvantages of both extremes.

Binary Divisions And Advanced Cognitive Evaluations: Discussion

The discussion in this and the previous three sections is not meant to suggest that the human mind necessarily works in an exact fixed binary mode which is absolutely pre-determined. The specific nature of pre-knowledge guided binary divisions as developed by a given individual vary depending on the power of: generating stimuli, past memories, degrees of development of other knowledge units, and so on. In the brain neurological impulses cover relatively great distances searching for clues and arranging and re-arranging binary divisions, before formulating knowledge units (often as images). Sometimes only a few binary divisions are conceptualized before an emotionally charged holistic image emerges; at other times, vast numbers are involved before the same result is achieved. Indeed, it is the brain's ability to work continuously on a multitude of perceptions that makes it more than a computer (Cf., Greenfield, 1995) and renders strict binary fixities unlikely.

What is important is the tendency of human cognition to continuously split stimuli into a limited number of species typical binary divisions and recombine these into holistic units/images, so that regardless of the amount and length of analysis, a very limited number of consciously perceived variables exist at any one time. As a result, regardless of quite extensive environmental variation, human cognition reduces the world to a very limited number of evaluated variables and continuously works to keep the number limited. Moreover, images of these evaluations are hierarchically arranged and because, the extremes of binary oppositions often come to be perceived as dangerous, there is a tendency to cognitively and emotionally identify the middle of a number of intersecting continua as being normal and to conceptualize the extremes as being potentially anti-social/evil.

These processes are clearly extremely significant for understanding how human learning and thinking operate. This includes understanding the vitally important roles of daydreaming, fantasizing and thinking. But they also begin to show how the development of religious, political and scientific theories come to be a significant aspect of human consciousness. Equally important, they are very significant for understanding the way in which self-images, social goals, taboos and boundaries are created, defended and generally adhered to. The human mind, it seems, cannot cope with unlimited, random, indeterminate variables. The tendency to the reduction of variables means that humans never have to deal with more than a few at a time. The tendency to hierarchy means that the relative importance of each variable is constantly under review. Both mean that environments, physical and social, are quickly evaluated and often reacted to.

These processes have meant that regardless of how difficult or complicated environments have become, basic human motivations have remained relatively quick and efficient. For example, the sexual potential of another person is quickly evaluated, and possibly reacted to, be it in a society of twenty-five people or of one million people, whether in a hunting, gathering, farming or industrial environment. Emotional-cognitive processes (including advanced cognition) have the capacity, and propensity, to set out a world which is not only comprehensible, but live-able in.

DELINEATING

Stereotyping

The worlds of people and things, then, are often reduced to charged stereotypes. Families, for example, can become not so much the realities of moving moments between good times and bad, joy and sadness, affection and punishment, but rather a image of perfection, or of engulfment and repression, to be sought after or avoided at all costs. Employees can become stereotyped as two or three variable images on an organizational chart, bosses two or three types in the minds of workers. Capitalists become selfish, lying, greedy caricatures of humanity - capitalism itself being reduced to the dimensions of profit motive and class exploitation. Politics becomes a struggle among the 'right', moderates, and the 'left'. Foreign countries become friends, neutrals or enemies, each represented as one player in the game of international power politics.

All members of a group are given the same characteristics: For example, during the Cold War, all Russians became hard, tough, devious, lying atheists, epitomized by a caricature of Stalin, or else brave, heroic, altruistic models of the future, epitomized by a caricature of 'the honest worker' - instead of 350 million individuals caught up in oedipal struggles, love triangles, sexual frustrations, creative exuberances, friendships, money worries, spiritual confusions and so on. Immigrants become lazy, stupid, silly and childlike, or brave, suffering, persistent pioneers.

I could go on with examples but the notion of stereotyping is well known and generally well accepted by social scientists: what I wish to argue here is that stereotyping is absolutely normal. Cognitively it is the way humans bring their environments under control and render them understandable. There are positive stereotypes just as there are negative stereotypes. Each provides messages which contain a huge amount of very quickly 'analyzed' data. Reproductive and survival success depends not on knowledge about the infinitely complex universe, but upon motivators which draw us to a very few, well-chosen individuals, cause us to defend against some others, and motivate us to avoid, as far as possible, most others.

So when we walk among a crowd of strangers, each individual is treated as a potential friend, a potential enemy or, much more likely, as part of the mass. We avoid eye contact with members of the mass and, despite the density of many crowds, we also make efforts to avoid physical contact. And when we can't we take great pains to avoid any sort of suggestive (aggressive, sexual) contact. A mild smile and a deferential posture is all that is needed to place us in the stereotype of non-threatening neutral. A big smile, on the other hand, is potentially suggestive, just as a frown is potentially threatening. The crowd is a forest and we hope to blend in unseen. Our friends are few, our leaders are reduced, in our minds, to at most two or three. Even in democratic societies the complexity of governmental institutions are reduced to the name of the elected leader, and in other countries to an image based on one or two stereotypes of the people and/or of their leaders.

Stereotyping does not just take place among the 'masses' or the non-educated, but among all human beings. Leaders of states often see the world in terms of the opposition parties, two or three types of voter, 'the economy', a few other countries and/or regions of the world. To be sure, it is easy to split each of these into a number of dimensions, when concentrating on any one of them, but even here the variables are usually continuously reduced to charged holistic images in order to keep the number which an individual has to deal with small (to reduce both the ambivalence and powerlessness which the ambiguousness of detailed distinction can generate).

Identification

By the time social stereotypes are formed, offspring have already identified their mothers and a few others as being safe but a large part of the world of people as either to be avoided or ignored. Identification, however, is more than just designating types of individuals; it also includes a strong feeling of attachment of self to certain of them, to the point, in some cases, of almost merging self with specific others, or at least self's destiny with them. This leads to parental engulfment, matehood, mutual respect, awe of godfathers, hero worship, being a fanatical fan and powerful commitments to political leaders. When stereotyping identifies self with individuals who belong to specific groups, and these feelings are transferred to the groups as abstractions, it leads to such phenomena as the supporting of a team, fanatically following a band, advocating the ideas of a political party, consumer loyalty and nationalism. In so far as identification leads to more direct membership in groups or organizations, it results in such things as, for example, final piety, kinship loyalty, gang bonding, team loyalty, patron/client obligations/respect, acceptance of and honour granted to professions and, again, nationalism. Humans can also identify with the goals of the groups to which they are members or otherwise attracted (A. Freud, 1966; Badcock, 1986, 1994), resulting in identifications with shared abstract ideals.

With the process of identification with groups and abstractions it is often the case that individuals identify themselves more strongly with certain other individuals, especially leaders, within such groups than they do with the groups as abstractions. It is difficult to tell, also, the degree to which identifications with abstract goals only apply when: 1) they are highly charged at a very early age, 2) are continuously advocated by individuals who are loved, respected, or otherwise admired by self, 3) adherence provides some combination of status, sense of power and significant material rewards and/or 4) a group's abstract ideals become part of a process of rebuilding self after a major breakdown. In other words, we cannot consider the processes of identification as being separate from the human nature propensities which underpin behaviour generally; identification with God does not free humans from their bio-electric nature.

Whatever the exact nature of the causation, individuals do come to identify emotionally and cognitively with other individuals and groups and with certain abstract ideals put forward by these groups. When such identifications become highly charged they can easily become part of self-identity, that is, become addictive: as a result the claims of such groups most often become translated into needs, rights and purposes of self. As a result, members of groups can come to believe that the aims of their particular group represents the major means of providing satisfaction of the fundamental 'needs' of not only group members, but very often of humans in general - if not of humanity in the abstract. When identifications result in individuals and groups treating their aims as the purpose of human existence (and the aims of others come to be seen as a conspiracy of anti-purpose), we have the combined effects of a number of additional emotional-cognitive tendencies/propensities. These can be seen most clearly in human deterministic thinking, teleological reasoning and spirituality.

Determinism

In looking to explain their own and the behaviour of others, humans tend to attribute deterministic, causal forces to three major dimensions of 'humanness'. These forces are often treated as being potentially, if not in fact, uncontrollable. They are considered to be the causes of behaviour generally, and, as part of the stereotyping process, to cause all members of certain groups to be and act basically the same. The first dimension of humanness given causal power is biology. It is often felt that specific biological characteristics (e.g., genes, physical strength, stature, skin colour, facial characteristics, innate intelligence, hormones, inborn character traits) determine individual behaviour and, when shared by a group, make all its members think and behave in a similar fashion.

The second dimension is spiritual. It is assumed that there is a spiritual dimension to humanness (the soul) which is not only receptive to forces greater than the immediate biology of being, and of reason, but, indeed, unable to resist those forces. Here it is argued that specific individuals, or whole groups of individuals, can become infected (or at least greatly affected) by supra-human, if not supernatural, forces which determine individual behaviour and make members of groups similar to each other (for example, those saved by Divine Grace, God's chosen, the pure, the polluted, the possessed, witches, the mentally ill).

The third dimension of humanness given causal power is Will/ Rationality. Here it is believed that specific individuals, or entire groups, have degrees of will power and/or capacity for the use of rationality. This causes them, at the one extreme, to use reason, be wise, logical, strong of character and independent, or, at the other extreme, to be silly, illogical and easily brainwashed. Such people, it is often believed, tend to give blind obedience to all sorts of cultural and political values and beliefs.

All three of these dimensions of humanness provide the basis for human notions of causation, it can be argued, because during the development of self they are universally problematic, potentially dangerous and seem to have a power which is not easily controlled. As a result, knowledge units relating to them very often become highly charged with both hope and fear potential. It must be noted at the outset, however, that these three dimensions of humanness are, in practice, not as easily separable as the above definitions imply. There is considerable overlap in the way they work, and in the thoughts of a given individual it would often be impossible to clearly distinguish the dimensions of humanness they are attributing as the basis of causation in specific instances. As analytical categories, however, these distinctions provide a useful aid in looking at human approaches to causality.

Let us consider the biological dimension first. For a child, the texture and shape of a mother's (parent's) breasts, body and face, and the security of her embraces, become highly charged as safety. This addiction is a child's continuing umbilical cord in a world in which a biological cord no longer unrelentingly provides nourishment and security. Nonetheless, from time to time, this cord is, without reason from the baby's point of view, suddenly withdrawn. A strong sense of abandonment ensues, even when other textures, shapes and physical actions - strangers - are substituted (these have not received the same emotional-cognitive charging, they have not become addictive). It is easy, we can speculate, to project feelings of tension/anxiety/fear generated from ‘abandonment’ onto the apparent coldness, distance and seeming power of intruding strangers. In other words, it is easy to learn (consciously and unconsciously) that the perceived physical behaviour of the mother and stranger mean safety and danger respectively, and, as a result, that biological being has causal power of considerable significance And the message that ‘bad’ biology can be dangerous will never be completely overcome. Most children can, eventually, successfully come to terms with the fact that they will have to interact with a considerably greater number of individuals than their mothers, but few will ever have the same unguarded faith in physical intimacy (umbilical cords) again.

This is only the start. During growth, physical size itself is problematic for a child. Not only are bigger/older persons much larger than a growing child, they appear to be able to do whatever they want (such people swing children about as they, the intruders, will it; they physically prevent young children from doing things, they tease unmercifully). If an adult is hurting a child there is little the child can (consciously) do. Even when a bigger/older person protects a child, the growing child can only marvel at the physical power - the apparent capacity to do anything – a capacity they, themselves, clearly do not have. Will they ever be allowed to do those things?; will they every be able to do those things?; will they ever grow up? These question haunt; so the growing child tries something he/she shouldn't, and fails, and is punished, and is told how stupid that was, and the questions come back, and they now include, 'am I worthy of growing up?'; will something prevent it because I am not worthy? And there seems to be no escape from this dependency or danger - except in fantasy play, often with the aid of toys which give power to the child, such as dolls to be equally manipulated and toy guns to magically destroy the superior power.

As the child grows, during the development of the adult self, personal capacities increasingly tell us that we are not achieving everything we fantasized about, and assumed big people could do if they really wanted to and tried and worked hard; for example, we are not all that good at football and tennis and running and swimming and looking like a model. There seems to be a real possibility of an insufficient development of both primary and secondary sexual characteristics, especially of our breasts or penises. We masturbate like mad and nothing happens – will it ever?, have I been stopped at this point?– am I forever the baby child; there is a fear that we will not grow-up physically. (Persons too small, such as dwarfs, act as reminders that self may forever remain a child, powerless, silly, patronized, humiliated – Fiedler, 1981). There is the constant fear of being, and remaining, ugly, physically revolting, covered with pimples, scoured with spots.

Early on we emotionally learn of the potential dangerousness, and evil, of various body expulsions (for example, excreta, urine, sweat, mucus, pus, tears, vomit). These come from our body, we are rotting inside, but there is nothing we can do about it. These substances seem always to be spewing forth to keep us in a defiled state. Much later we begin to age and once again there is nothing we can do about it. Strength begins to fade, we become tired easily, wrinkles and aches appear, hair turns grey, or even falls out, we lose our youthful attractiveness, our vitality, our knowledge that we are going to change the world and achieve happiness. Our children move away and begin to avoid us. We begin to become children again, unable to feed or toilet ourselves; we hate it. We begin to slowly die, and, despite everything, we do not want that either.

Much of biology, thus, seems to be a powerful causative force, often out of our control. The dangers of our biology direct our daydreaming and thinking towards willing the normal, natural, for self. As a result of attempting to hang on to a degree of normality, however, individuals of noticeable different stature, skin colour and/or facial appearance come to be seen as unnatural, and to be so because of uncontrollable biological causes. If these causes affect physical characteristics might they not also affect, if not control, behaviour? - after all, they generate ill feeling in and sometimes desperate behaviours on the part of self And if this is the case, might not such people themselves be dangerously out of control? Thus, as a result of projecting self's own anxieties about the potentially uncontrollable nature of biological processes, characteristics and appearances, different biological images most often come to include, during stereotyping, a number of threatening behavioural attributes. So, strange others become people not just with 'strange' features, but also with such characteristics as extraordinarily large sexual appetites, uncontrolled aggression; they become people who are cold, uncaring, vicious, low in intelligence, in short, not quite human.

Now let us consider the spiritual approach to stereotyping. We have seen that a fear of our own dangerously uncontrollable emotions, and some of our desires, often attracts us to somewhere near the middle position in our cognitive hierarchies. This cause of willing the normal (average) generates restrictions on our more basic emotions as well as on many of our desires and fantasies. It is enforced by conflicting, charged, bio-electric impulses taking the patterns of what I have described in Chapter Five as guilt, the fear of showing envy, shame and embarrassment, for example. In the process, there seems to be very powerful forces at work which are not altogether under an individual's control. Certain emotions continually creep into our fantasies despite their prohibition and the sense of danger they engender; for example, lust for persons other than the one self loves and to whom self has promised fidelity. Or feelings of envy and jealousy continually invade our consciousness despite strenuous efforts to keep them submerged, indeed, even to deny their existence. And when depression hits, it often does so intensely, to a point that it seems to be completely out of self's control; its reasons for existence often seem to deny rational explanation.

Even at the stage where danger feelings have been turned into more cognitively organized fears and desires (and where generally accepted taboos, rules, avoidances and boundaries have been established to help individuals avoid dangers), frantic-feedbacks generated from conflicts between and within fears and desires can still make it seem that we are imbued with inner forces over which we have little, if any, control. We get extremely excited when we score well on a test, and when our team wins, despite not wanting to gloat or to look like a silly sports fanatic; and we get very nervous when we go for an important interview, even though we know we have the best qualifications by far for the job; and our desire to be loved and admired, and for status or to be in control, or the fear of losing face or honour, for example, all seem to be there despite our efforts to banish, or at least control, them. And sometimes we cheat on our tax returns and lie about our achievements (not to speak of secretly – probably only half-consciously - trying to steal our friend's lover, or desiring our buddy's wife) and then feel silly and guilty and very much afraid of being found out, but we seem not to have been able to stop ourselves; and we doubt we will be able to in the future. There seems to have been a force for bad which has got inside us, driving us. Those who do seem to be in control of their inner soul, on the other hand, seem to be imbued with some sort of power of goodness. Whatever our exact experiences, forces greater than ourselves, and others, and seemingly above and beyond biology, appear to be in control of our destinies.

Humans have devised means of trying to deal with such spiritual dangers (and opportunities) as outlined above, and in the process to also harness the powers of 'Good'. These means start by developing notions such as saintliness, depravity, good, evil, purity, pollution, grace, disgrace, holy, unholy, for example, so that we get to grips with danger through naming. At the same time, procedures for drawing Good to self, are established. So those individuals who regularly undertake purification rites, pray a lot, condemn the Devil (and his followers – unwitting as they might be), strictly adhere to accepted moral doctrines, present self in terms of completely suppressed emotions and desires are felt to be good, pure, in a state of grace and moral, while those who do not are considered to be polluted, cursed, in the grip of a sinful soul - and to be immoral.

This common approach to causation clearly allows for a hierarchical arranging of individuals in a community in terms of their perceived value (to oneself and to the community). It encourages the stigmatizing of certain behaviours (and individuals) as being beyond the pale, and others to be sought after, and befriended, with great enthusiasm. If all (or almost all) members of a group are perceived to be in a state of grace, or, on the other hand, cursed, for example, a social stereotype comes to exist. From there it is easy to attribute to a religious, ethnic or national group a set of characteristics. The stereotype of them can even come to include the idea that their members go beyond the normal expression of dangerous emotions because they encourage and approve of living by such emotions. So, it may be believed that they indulge in abnormal sexual practices - they are at it constantly or never, they start at the age of five, do it with animals; all members are inherently devious with a tendency to lie, cheat and steal; they are greedy, untrustworthy, wild, unpredictable, unreliable, and so forth. It is considered that their unhealthy spirituality means that their emotional impulses are fully in control of them and are tending to run wild. And even worse, they seem to like it and are not intent on doing anything about it.

The third category of causes (Will/Rationality) attributed to individual behaviour and to stereotyped groups leads on from the above. It is related to the human tendency to judge ourselves and others in terms of the human ability to rationally control ourselves and to be able to defer gratification. That is, most people, at least in modern times, believe that most of our emotions and desires are not as uncontrollable as the above discussion suggests. Emotions, desires and fears, it is felt, can, in fact, be made subject to rational control relatively easily, if there is a will to do so. Besides encouraging self-control, humans establish rules, goals, taboos, procedures, hierarchies and cognitive boundaries to guide us and to provide predictability in order to allow rationality its full reign.

This belief in the power of rational control, however, does not diminish the human propensity to deterministic thinking or to stereotyping; indeed, often the opposite is the case. Being able (or not) to adhere to rules, taboos, hierarchies and boundaries, to control emotions generally and to defer gratification, comes to imply a strength (or weakness) of will and rationality among individuals and/or members of a group being stereotyped. Individuals are said to be either clever and strong of character - imbued with "The Right Stuff" (Wolfe, 1980) - or rationally incapable and/or too lazy to exercise human will. The second type, it is felt, do not wish to control rampant, selfish, emotionally-dictated desires. They are said to be superstitious, pre-modern, easy prey to such things as charismatic demagoguery, the mass media, foreign ideologies, oratory and mass culture.

The belief in this type of causation is commonly found among members of elitist groups (such as existing aristocrats, or socially selected or self-selected candidates for implementing ultimate truth). Such exclusive groups (real or imagined) are themselves easy victims of the stereotyping process and are thus quick to use it in defense. Very often such groups stereotype themselves as being clever, strong, righteous, dedicated and misunderstood, and everyone else - except their conspiring enemies - as not so clever, lazy, deluded and generally without minds of their own ('the masses'). Often, however, a conspiracy element is strong in this pattern of stereotyping, so the opposition are pictured as being cleverly devious rather than stupid. This may be because elites often demand complete compliance/ agreement and/or co-operation. But since such aspirations never work out as smoothly as desired, elitist goals are seldom realized; it is therefore easy for obsessed individuals to rationalize their own failure and the seeming success of certain other groups on the basis of a postulated fanaticism and robot-like obedience among members of such other groups, masterminded by an extremely clever conspiracy.

Throughout modern history, for example, many people have believed that Jesuits brainwash children in order to completely destroy individual will for the purpose of eventual world conquest (spiritual if not physical). Freemasons are believed to contrive to get rich at all costs, to protect each other from the law and to manipulate governments. The mythological ruling class has, according to some, no end of extremely subtle means at its disposal in its drive to destroy the will of the people and thus sap their potential discontent. Governmental elites are believed to spend all their time conspiring to lie and deceive voters. Governmental elites, in turn, are convinced that they face a number of conspiracies, from the media, academics, trade unions and business groups, not to mention from their own civil servants, opposition parties and other governments; everyone, they feel, seems to have as a goal, if not life's purpose, their, the elite's, undoing.

Teleological Thinking

With the notion of purpose, however, we come to another important dimension of human cognition, teleological thinking. Besides attributing causal powers to certain felt to be often uncontrollable biological, spiritual and will/'rationality' forces, humans have a tendency to generate holistic images in which the individuals and groups they encounter are part of a larger scheme of things. This larger scheme most often is felt to have its own 'essence' and purpose, both usually believed to be pre-designed. This easily grows into the idea that such an essence and purpose have considerable powers of causation in and of themselves; that is, the larger scheme is attributed with power to fulfil its own pre-determined destiny; it becomes what Nietzsche (1990) has called, the 'cause-in-itself'.

At a somewhat less cosmic level, designed essence and purpose are sometimes represented as having structural or organizational characteristics, especially when humans are trying to theorize the conditions of social order or solve major social problems - the achieving of which are felt to be necessary to fulfil the human purpose. For example, social structures, and sometimes organizations, have been given a greater than human essence and, thus, have become 'cause-in-themselves' by being considered to have needs which must be fulfilled. Such necessity, thus, becomes the cause of the parts of the wholes. Families as sub-structures, for example, are sometimes said to be necessary for social stability which, it is argued, is necessary for the continuation of the structure of society, and society, defined as a social structure, is necessary in order to provide a secure environment for families. Another version might be: families are necessary for reproduction which is necessary for the continuation of the species - treated as a holistic entity with a spiritual destiny - and society is necessary for the protection of families.

None of these, of course, is a need or necessity in any absolute sense. At this level of abstraction, however, the teleological notions of design, purpose and need replace such things as emotions, desires, fears and personal expectations as causes of human behaviour. This form of teleology and circular reasoning is typical of the highest levels of human cognitive thinking and has been a favourite of theologians, philosophers, social scientist and many biologists alike when they have set out to explain the essence of the human condition, especially human social life and history. Teleological thinking underpins not only beliefs about families in relationship to societies, and fantasies about them, but also human conceptualizations of more formal organizations, from relatively small scale organizations to nation states. Teleological reasoning plays a major role in the development of universalistic social theories and religions. It is the ultimate in reducing a multitude of conflicting, confusing variables into a smaller number of comprehensible, controlled, meaningful and 'purposeful' variables.

Teleological thinking is not, however, strong enough, or comprehensive enough, to overcome basic human emotions, desires and fears. As a result, one person's purpose, destiny or needs are another person's alienation, hell or curse. And, more importantly, the satisfaction of one person's purpose, destiny or needs is often a danger to someone else. Furthermore, purpose, destiny and needs never remain static. Like the desires from which they are unconsciously drawn - and contrary to what the words usually imply - these notions continually change and often escalate. Moreover, as noted, as far as we can ascertain there is no need for anything. For example, we do not need to eat because there is no need for us to survive, or for the universe to exist, for that matter.

The concept of need is a reified product of emotional-cognitive charged human teleological reasoning in which we have deceived ourselves into thinking that there is a real purpose and destiny to our existence. This cognitive process has clear evolutionary advantages for a self-aware creature in that it generally motivates survival and reproductive behaviours without philosophical prevarication. As a by-product, through giving meaning to existence, truth, within that meaning system, can exist. To seek truth in this sense is to seek to understand the meanings specific humans, and/or groups of humans, attribute to various phenomena, and thus to better understand self in relationship to specific social others.

In evolutionary terms, again, understanding self in relationship to others most likely was a reproductive and survival advantage, and we can speculate that selection provided both the abilities and desires for seeking truth in the above teleological sense as a major motivation for gaining advantages over others. To seek truth clearly can be a manifestation of the human 'will to power' (Nietzsche, 1990). However, this would have meant that those who could fake truth would have an advantage because it would throw others off the track of competed for goals, and so we can further speculate that selection (especially in a self-aware creature) also favoured means of presenting a false picture as truth.

DECEPTION AND SELF-DECEPTION

In fact, in a number of species of plants and animals, messages that in effect deceive others are common (Cf., Trivers, 1985; Hauser, 1997) Among a number of animal species, for example, aggressive bluff is frequent in sexual, and other, competitions. From the perspective of natural selection, the original aggression most likely would have resulted in real violence, but selection would have provided mechanisms for it being defused by additional stimuli which cut-off the aggression at a point when opposition began to back away (Eibl-Eibesfeldt, 1971; Gilbert, 1984) because even the winner is likely to die from wounds (infections) in real violence. With a self-aware creature this instinctive bluffing has to a considerable extent been replaced by conscious, cognitive bluffing.

Human beings alter their appearance and hide defects in order to attract others; humans present expressions and gestures to convince others of feelings and intentions different from the ones they really have. It is common for people to leave out information, distort information and to simply lie. Human language serves well here. If language evolved to convey the truth, or even just information, it would be considerably simpler than it is (Badcock, 1986). A few simple codes would do. But language is full of redundancies which convey elaborated images rather than pure information, and it also includes considerable "noise" which can obscure almost anything (Campbell, 1984), especially an individual's intentions.

In service of this, gossip is an extremely important mechanism (Cf. Dunbar, 1996; Pinker, 1998). Gossip allows us to find out what is going on and to believe that we are at least as good, if not better, than most of those we are suppose to trust. During gossip we and our co-gossips reinforce each other in the conviction that we are loved, good, worthy and generally respected while the objects of our gossiping are not so good or worthy. During gossip we conspire with our co-gossips to accept each others' delusions as truths. During gossip we get others to conspire with our desire to harm those who have transgressed against us, we take joint verbal and imaginary vengeance on them and together conclude that they have done wrong.

The same applies in so-called serious talk. During meetings among professionals, for example, individuals not only convey what they want conveyed and hold back what they want to keep secret, but also conspire to accept each others' 'masks', delusions and hopes. They often conspire against outsiders, dressing it up with the language of merit, commitment, duty and loyalty. This is especially the case during political meetings where careers, factional positions, the role of specific political parties and the nature of the polity itself may be at stake. Indeed, it is in the political arena, perhaps, where most deception is practised. (Its possible competitors are to be found in sexual and family relationships.) This is because it is in politics that social truths and falsehoods are established; it is here that humans place a significant degree of their trust for a better future. It is in politics that 'power' often has no 'big brother' to curb its excesses or to dictate truth - or if it does, its truth is even more distorted than that of the average politician's.

From an evolutionary point of view all of this is quite understandable. Humans may be in competition with each other for mates and for being in control of the effects of the circumstances in which they find themselves, but it is often the case that for the best success they should not be seen to be so. For example, when an individual is trying to get a potential mate to contribute to the welfare of future offspring on a grand scale, while self searches for better genes for the offspring, or self seeks to spread genes as far and wide as possible and provide no childcare whatever, it is not in their best interests to let this be known to potential mates. At the same time, it is in an individual's interests to find out about such deception and to receive support from those trusted when becoming a victim of deceit. In this regard the obviously extensive human capacity to rationalize, exaggerate and reinterpret, in order to regain the self-respect, confidence and loyalty of friends - crucial to continuing reproductive success - makes senses. Again, the role of gossip looms large.

Thus, during gossip and serious talk ". . . All the marvels of cognition and information manipulation that language permits can be used in the service of self-deception" (Trivers, 1985, p., 416). A salient point concerning language is that each of the above methods of self-deception is considerably more convincing to ourselves and socially effective if we have verbalized it a number of times and our friends have signalled verbal, and non-verbal, acceptance. "Needless to say, no one who shares a delusion recognises it as such" (Freud, 1957 p., 36). And when it becomes part of folklore, official minutes or statute law, it is no longer a delusion, or at least few dare call it such; and, anyway, if they do, they only have other delusions to offer.

Nevertheless, as noted, if selection favoured those who could deceive it also favoured those who could spot deception and use it to their own advantage, or at least see to the moral condemnation of the deceivers. Few people are completely convinced by gossip or by what spokespersons for large organizations or politicians say. Many do not altogether trust even their best friends. Language as a means of deception and for sharing delusions, significant and creative as it can be, is far from perfect. Lie detectors, perceptive individuals, events and history, for example, can relatively easily uncover verbal and written lies. Even a lie detector, however, can be beaten by a self-deceived individual. There has thus been selection for a strong capacity for self-deception in order to successfully carry out deception in regard to others, including ". . . hiding the truth from the conscious mind the better to hide it from others" (Trivers, 1985, p., 415). This characteristic of evolution is, it can be argued, especially significant for our understanding of human behaviour (Badcock, 1986). Much of human cognition results in unconscious self-deception. Most of the stereotyping characteristics of human cognition, for example, contains, from a scientific point of view, in-built deceptions. As we have seen, a multitude of potentially conflicting, most often extremely complicated, objects, behaviours and individuals are reduced, stereotyped and explained in terms of relatively simple images (images with teleological causal elements neatly built in) to then be presented as truth.

Robert Trivers (1985, pp., 418-420) has identified a number of additional ways in which cognitive self-deception works. The first is "beneffectance" where we convince ourselves that we are beneficial and effecttive. We deny responsibility when something undesirable happens, and exaggerate our role when something desirable occurs. The second is exaggeration itself. We exaggerate our good features, acts and feelings and downplay our less desirable ones. A version of this is to invent experiences which place us in a good light – we were there when it happened, when the prophet reveled the truth. The third method is "the illusion of consistency" where we selectively re-write the past in order to justify the present or future plans. The final method has to do with differential perceptions of relationships and events in such a way that we see ourselves as right and good, and others as not so right and good. We ignore, cannot concentrate upon, cannot lock in as permanent knowledge units, many things with negative personal connotations - psychologists have called this cognitive dissidence.

Human cognition, then, contains a great deal of rationalization. This characteristic of cognition has made countless individuals relatively successful participants in reproductive and survival competitions. It also makes the world comprehensible for humans. But more importantly, it motivates a highly self-aware creature into thinking that there is a purpose to its life and to the existence of the universe. Otherwise, it would be true that even 'a little deep think-in drives you to drink-in' (as Willie Nelson and Merle Haggard point out in their song, *It's My Lazy Day*). Those who think there are five sides to every question are not usually good parents, workers or politicians, or much else for that matter. Self-awareness opens the door to considerable, potentially destructive, self-criticism. Self-delusion has, for most of the people, most of the time, prevented this. It may be one of the reasons why evolution has not eliminated the subconscious (internally charged knowledge units which do not, or only very rarely, surface as clear images); too many reproductively important motivations could be severely damaged if exposed to conscious analysis.

One can also argue that it is for this reason that natural selection has not eliminated human spirituality. Human spirituality is the human tendency to get 'high', to be lifted, by positively charged holistic images of a glorious future state of being, of a higher human purpose, of eternal bliss where fear is absent, depression gone, desires fulfilled. Brain opiates flow and sensitivity reaches a pitch of trembling excitement when individuals focus on the symbols and people visualized in such images. Human spirituality creates a land of synchronic harmony where tranquillity has its own teleological reasons for existing. The human unconscious and human spirituality have made possible a considerable degree of sociability for humans. The more people delude themselves that they are altruistic the more likely they are to act in a co-operative and altruistic manner (Badcock, 1986); the more individuals become drunk with a higher purpose, the more likely they will live for it, and consequently, for others.

SUMMARY AND CONCLUSIONS

A number of species typical cognitive processes have been identified in order to outline the manner in which human nature organizes various stimuli while charging certain of them with emotional-cognitive force. These processes evaluate and quickly convert a number of varied and often complex stimuli into comprehensible information. They also generate powerful notions of 'the truth' of a number of reproductively advantageous ideas and behaviours so that social action in regard to them is decisive, purposeful and unhesitating. This process also includes the development of very strong social attractions and repulsions which in turn motivate the establishment of social goals (desires perceived as needs) and restrictions (in the forms of taboos, rules, social boundaries, social hierarchies and political formations) which go to make up species typical patterns of behaviour. In order to further our understanding of how this works, let us consider in more detail some of the human nature propensities/tendencies which lead to various taboos, rules, boundaries, hierarchies and political formations.

CHAPTER SEVEN
HUMAN NATURE AND SOCIAL PROPENSITIES

The argument so far has been that, in order to avoid the grand circularities and teleologies of much of the theorizing and systems-building during the development of the social sciences, a notion of human nature has to be developed. This is to give a social science a basis in reality – in biology and evolutionary theory - rather than simply being based on un-substantiated enlightenment-inspired notions of direction, purpose and progress. The model of human nature which we require must, in other words, not only remain consistent with the purposeless and directionless nature of Darwinian principles of evolution, but also incorporate an understanding of the amoral principles of genetics and the bio-electrics of current neural sciences. It must be a human nature which is not just a black box out of which magically spring all the ingredients for humanness – be they goodness, perfectibility and potential happiness or, conversely, evilness, imperfection and misery. It has to be a human nature that can realistically be seen to contain *motivators* of human behaviour, for better or for worse.

In this work I have set out to outline such a model of human nature. The result has been a picture of human nature as a number of emotional-cognitive processes (including advanced cognition) which do not make up a perfectly designed (magnificently adapted) engine for creating synchronic, stable social systems or structures. Human nature is not what it is because it has been 'created' in order to provide human perfectibility, happiness, or anything else for that matter. Rather, human nature is what it is because of its evolutionary history, a history in which natural and sexual selection piled new mutations on what already existed in a constant 'struggle' for reproductive success This has been, and is, a struggle in which males and females do not have identical interests; children and parents not uncommonly manifest conflicting demands; males compete with each other, females vie with other females; families find themselves at odds, and so on (see note, four p., 234).

To complicate matters, during the evolution of the emotional-cognitive mechanisms which developed with these reproductive conflicts, humans became aware of their own independent identities and of their sexual identities. So, self-awareness evolved in a context full of potential dangers. Evolutionary theory (along with anthropology and history) suggest that these dangers included possibilities of: parental abandonment, sibling bullying, sexual rejections, romantic humiliations, cuckoldry, misplaced friendships and reproductive exclusion (see note four p., 234). During evolution, humans developed species typical emotional - cognitive motivators, desires and fears which gave self a degree of protection (and thus reproductive success) in the context of these dangers.

So the lusts, loves, envies and jealousies so common in today's 'battle of the sexes', and the anxieties, angers, loves, devotions, hatreds, worries, feelings of guilt and senses of despair (for example), so often expressed in relationships between generations, friends, enemies, colleagues and kinsfolk are ingrained (as motivators) in human nature. As a result, in protecting self there is a great deal of attention seeking, cajoling care, competing for mates, attempting sexual/romantic conquests and claims for social recognition/honour. In the process we make friends and enemies, bond and split, seek vengeance and forgive, form alliances and defect, support kin and fight kin. We submit to and escape from powerful patrons and we distance ourselves from certain others and manoeuvre for advantage, protection and privilege within the larger community. In these activities self is often on-the-line to impress and to live up to obligations. As a result, selection seems to have favoured a relatively elaborate capacity for self-*image*, which includes our expectations concerning the care and recognition we desire (feel we deserve), as well as our perception of a safe position, for *us*, between extreme independence (isolation) and engulfment (a complete lack of individuality). But it is a human nature in which, nevertheless, there is a constant awareness of potential failure.

It seems, then, that our evolutionary history has left us with a master ontological conflict within human nature; a conflict between those emotional/cognitive processes, desires and fears which motivate independence and individuality (the human will to power) and those which motivate submission and dependency (human sociability) – together they make up 'the human condition one might say. This condition, or dilemma, has, in fact, long been known about by numerous philosophers and great storytellers. But neither these, nor celebrated theologians, creators of utopias, political messiahs or scientists have been able to eliminate it.

Still, humans do come together, and they do work together; they do form alliances; they do love and rear offspring. This too is confirmed by history. So the question has to be, how can both of these types of behaviour – antagonistic and anxiety provoking and loving and security providing - come from the same human nature?; not, how can we fundamentally change this human nature or get rid of it altogether?. To their credit, in fact, a number of the great thinkers of the past have not tried to completely resolve the dilemma of human nature, but rather to make up exciting stories about it and to learn from it and to seek to live with it. And from their work (and from history) there appears to be an acceptance of another legacy of the human dilemma, a legacy of perpetual politics. Poets and playwright, composers and philosophers, chroniclers and preachers - when not concerned with sex, lust, love and romance - have been fascinated by the intrigues of politics. Stories of conspiracies, deceits, displays, marriages, adulteries, heroics, victories and executions among those who have played the 'who is to be a hero?'; the 'who is to be a court favourite?'; or, indeed, 'who is to be king?' games, are the stuff of their work.

And rightly so. If we take politics to be: enticing, seducing, negotiating, persuading, forcing, judging, leading and following, for example, then political processes (politics), as politics with both a small p and a large P, are clearly extremely important in human affairs. Such politics can be seen at every level of human experience, from self to the state. The argument might therefore be, that politics play a significant part in generating, maintaining and changing the parameters of human social behaviour, at all levels. In other words, it can be argued that, somehow, human political processes have been, and are, the causes of sufficient protection, stability, predictability and safety so that we have been able to reproduce with great fecundity. This is despite humans having much less physical strength than competing apes and most other large mammals, and despite giving birth to extremely vulnerable offspring.

And it is also despite a conflict-prone and ambiguous human nature which can result in infanticide, murder, genocide and war just as it can in love, devotion, reasoning and mutual obligation. While human politics have arguably contributed to a degree of stability, predictability and safety, they also have been sufficiently flexible so that we have been able to adjust social behaviour to be able to survive and reproduce in a relatively wide variety of physical, social, demographic, technological and economic environments. Indeed, humans have been able to survive (and thrive in) relatively dramatic changes in all of these; we have even been motivated to *create change* itself, and to be sufficiently proud of what has been produced to call it 'civilization'. This argument would be in contrast to the view that politics only exists because of current faults and failures (albeit curable with right reason) in potentially rational - and perfectible - forms of social behaviour (again from self to the state).

HUMAN NATURE AND THE PROCESSES OF POLITICAL SOCIETY

The first of these views is very much the argument of this work. This is because, not only does it accord with a number of principles of natural and sexual selection concerning human reproduction and the evolution of human consciousness, (see note four, p., 234) and seems to be supported by vast historical evidence, but it also allows us to develop a non-teleological picture of social behaviour. If we can identify the active interpersonal *processes* (politics as defined above) through which conglomerations of emotional-cognitive processes, desires and fears act as *propensities* which result in *natural kinds* of behaviour, we do not have to rely on *design*, purpose or 'need' as causal forces in our search for patterns in human social behaviour. Of course, these processes must be based on a well established understanding of universally understood processes of natural and sexual selection, on evidence of what happened during human evolution, on what we know about brain functioning and on an understanding of human history (rather than on postulated rational/idealistic outcomes).

It is on this basis that I have been endeavouring to develop a model of human nature. To build on this, in this chapter I will consider a number of possible human nature *propensities* which can be argued to underpin different *natural kinds of behaviour* (specific conglomerations/mixtures of various propensities), which in turn can be argued to make up 'society' (and its parts); societies being conceptualized in this view as politically produced and maintained complexes. The contention will be that we politic to seek desire fulfilment and safety within the ambiguities and dangers generated by conflicting human nature, by the ambiguities generated *by our very efforts to deal with these dangers and conflicts*, and by the often unpredictable changing environmental circumstances in which we find ourselves. We endeavour to bring the dangers inherent within our environments under control; but that is not always easy; events are most often difficult to predict or to manage; powerful addictions, desires and fears are stimulated and enter into frantic- feedbacks of increasingly intense emotional, desire and fear reactions. And so there is a lot of running just to stay even; but where *is* even, how do we know when we are there?; that is, how do we know when we are safe? On the one hand, we are motivated to submit: to parents, to mates, to friends, to powerful patrons, to higher purposes, but at the same time to be in control, to dominate - at least self's condition. This conflict will not go away, and it manifests itself in every aspect of human social life

Yet, as noted, we have thrived. We have been incredibly successful reproductively, with time left over to build pyramids, cathedrals and go to the moon. And despite several attempts to eliminate groups of each other, we have filled up the earth, in some places so densely that the primatologists from Mars might well consider this to be the main characteristic for separating humans from other large primates. But this success clearly is not dependent on a synchronic human nature or perfectly designed social systems. It is for this reason – along with the ubiquity of politics in all types of human societies – that, as noted, it can be suspected that human politics themselves are not only natural to humans, but that their *characteristics* must have played a major part in so many humans having had considerable reproductive success.

Politics, then, as driven by conflicting species typical emotional-cognitive processes and species typical desires and fears, it is being argued, make up the very processes which result in human 'societies' and social evolution. To illustrate this possibility more specifically, and to begin to establish evidence for it, let us take the existence of a seemingly universal human nature propensity for seeking vengeance (Lopreato, 1984; Young, 1996; Kerrigan, 1996; Nietzsche, 1886) as an example of how potentially escalating and conflicting emotional-cognitive processes and desires and fears impact on human social behaviour, creating not only a politics of conflict (and change) but also a politics of social order and justice.

THE SEARCH FOR VENGEANCE AND THE SOCIAL ORDER

When others hurt or humiliate us, or our family or friends, we might strike out at them, or we might run from them. These reactions are born of emotional-cognitive reactions such as surprise, anger and fright. Such reactions are quick, but in evolutionary terms, perhaps too quick. If our adversaries are able to hurt or humiliate us, they may well be able to defeat us and striking out at them may get us into even more danger – this may be the very reaction they want as an excuse to do us real damage, if not take our belongings or displace us from our social position. If, on the other hand, we run we may get ourselves out of immediate danger but leave our loved ones in even more danger. More importantly, perhaps, we leave our possessions, and possibly even our social positions, to the attacker without even a fight. Even if our attacker is not interested in these, by running we lose standing in the eyes of our significant others, and in the community – we become a coward.

We have given up and may have to subordinate to the aggressor for no gain to self; running also demonstrates weakness and so others may take advantage of us. Our seething anger and sense of humiliation may be such that we later follow up by stealing back to take vengeance in a sneaky sort of way; but that further labels us a coward - including in our own eyes – very likely inducing a sense of guilt concerning our own inadequacies and the actions we have resorted to. In the eyes of the community we become devious, dangerous. We may try to prevent this by striking out at scapegoats – to try to demonstrate a lack of fear and to show bravado. But this has the real possibility of only alienating potential allies and/or getting us further labelled as being dangerous - out of control, and certainly not trustworthy; whatever the case, we remain both an emotional victim and undesirable in the eyes of the community; we lose both ways.

So, it can be argued, we have evolved the tendency to strike out (or restrain the other) when we are relatively confident that we are stronger, and to retreat with shouts of 'you bully, what was that for?' and, 'I'm *really* going to get you', when they are potentially stronger and/or the difference is not clear. In the second case, we *depend* upon our friends to join in the 'you bully' chorus and to make threatening sounds and gestures, and perhaps even to *restrain* us from directly attacking. We are protected from the aggressor by a show of solidarity from our friends, and if they restrain us, they have shown a concern for our welfare. We have the initiative and we do not look a coward. The aggressor's friends might show up and there will be a lot of shouting and accusing, and everyone will be a hero in their threats and reminiscences of the event.

All sides may eventually retreat and this may be the end of it. After all there have been a lot of heroes created. On the other hand, one side or another may still feel aggrieved. We might, for example, move away to consider *vengeance* with our friends. We may get fired-up and decide to take relatively immediate vengeance. This could work well through the element of surprise and the amount of group aggression we generate through frantic-feedbacks common in gang warfare. It may, however, set in motion a scenario which leads to an escalating taste for violence and heroics which take us well beyond a search for vengeance (or security). The result, for example, could be continual blood feuding with greater and greater destruction for its own sake, resulting in less and less individual security.

Equally, perhaps, it could lead to conquest by one over the other – which might provide a degree of security *within* the dominion of the blood brotherhood subdued surroundings. However, the eventual result of this could be perpetual rebellions, warfare, and, eventually, total destruction of the 'winners' because of massive alliance against them. Nevertheless, for young men, in many circumstances, vengeance groups can protect lone individuals in that if others know that they have potential mates willing and able to take vengeance for transgression against them they will be left alone. It also allows for the expression of a number of male desires to show off, to act brave, to bond with other males and to capture females and wealth – to be a hero.

However, theoretical consideration (and human history) seem to suggest that this is not an evolutionary scenario which often settles into a harmonious balance. Quite apart from its precariousness in terms of the reliability of human companions, the human capacity to kill - with sophisticated tools and deceit when it is felt necessary – means that any balance achieved through blood feuding is likely to achieve more blood than safety (or even heroic feuding for that matter). Moreover, it can be argued that, in significant part, human competitions for reproductive success have depended on much more than expressions of bravado and the exercising of vengeance, or even being of the warrior class. Humans are motivated to sexual showing-off and to sexual manoeuvring in a wide variety of ways. Behaviours such as decorating self, achieving material success, demonstrating caring traits, presenting self as: 'I am a poor country boy, come mother me', settling down at home, not to speak of: *femme fatale-ness*, adultery, cuckoldry and deviousness with acquaintances, and even friends, for example, are more or less common in what has been relatively successful human reproductive behaviour. All of these can put a great deal of strain on the obligations of total loyalty, bravado and complete dedication to the heroics of battle (and to each other) expected of members of blood brotherhoods. These strains are especially likely as members become older and enticed to settle down with a reproductive mate and have offspring.

And anyway, the above scenarios for seeking redress represent young men's games – and strong young men's at that; it is not clear as to its applicability to anyone else. (It is an approach which primarily serves male not female, reproductive interests.) Certainly, human history suggests that while a human desire for vengeance (and the 'necessity' for honourable satisfaction) is very easy to learn and remains strong, very early on humans began to feel and think that there were better ways to obtain security and justice (Cf., Aeschylus, *The Oresteian Trilogy* - Vellacott, 1956; *The Bible; Njal's Saga* - Magnusson and Palsson, 1960; Shakespeare, *Hamlet*; Mair, 1962). In some circumstances, for example, the formation of patron/client relationships, or some other form of relatively 'consensual' political procedures to provide security and justice, and a context in which honour could be maintained, evolved out of the dangers of feuding (see below).

But there are also other possible mechanisms which may have prevented, or at least stopped relatively quickly, tendencies to direct vengeance and blood feuding. The taking of actual vengeance, even with allies, may, in fact, have been restrained from the start by strong doses of bio-electric danger feelings (fear). If we do not strike out immediately these feelings may actually increase when we later plan vengeance. Or, we and our allies may take vengeance but suddenly develop a degree of sympathy for and/or empathy with the innocent victims (for example, the relatives of those we have harmed), generating feelings of guilt for a vicious and malicious orgy of revenge. These may be especially strong among those of our allies who were not hurt or humiliated in the first place. At the same time, our capacity for empathy and reasoning may well inform us of the dangers of the retaliation and dangerous escalations of violence which we must now contemplate. So, instead of striking out or taking physical vengeance, we may try to take a 'rationalistic' approach and demand compensation.

All of these feelings, behaviours and their verbal and symbolic expressions, may win us a certain degree of moral admiration, but they, nevertheless, run the risk of continuing to leave us subject to physical danger and of not completely getting rid of our status as victim(s). If we do not strike because of fear, we will never completely escape the tyranny of gossip in which we are deemed to be, deep down, cowards; we may wonder ourselves; shame and embarrassment will dog us. At the same time, feelings of guilt, sympathy and empathy, as a result of having taken vicious frantic-feedback inspired vengeance, may generate moral dilemmas - but showing them too much can result in others thinking that we do not know our own mind, that we are not really in control of ourselves and/or our actions – we are dithery, indecisive, weak; we make excuses; once again we remain victims and may be isolated if not picked on. And demanding compensation is useless unless there is some means of collecting it without losing even more face through failed attempts to do so; and anyway, compensation is often considered to be dishonourable blood money.

But here is another possible scenario. As we have seen, when we retreat with cries of 'you bully, I will get you', and our friends are in full clamour in our support, the tables are evened-up somewhat, for the moment at least. If together we go away and plan a quick attack and carry it out, the scenario described above pertains. However, if 'our cause' seems to have a lot of friends, and more and more join, and, as a group, we declare that we will be taking vengeance *later*, after some planning and consultation, the opposition has something to think about. And if we spend considerable time doing so, and in gaining public support, we can gain the high ground and become not victim but potential 'legitimate' enforcers of *acceptable* behaviour. So, in 'delayed vengeance', instead of taking the role of victim we, in our imaginations if not completely in public opinion, become the champions of justice and the advocates of morality. Those with whom we are close enough to confide in will (generally) verbally and emotionally support us in our mental vengeance and moral elaborations. The more people we can get to join in this support, the stronger our case becomes, of course, and the stronger our position. So, in fantasies, daydreams, discussions, pronouncements and planning about 'the need for vengeance', but later – perhaps in a 'fairer' way – we become heroes, heroes set to do right in the light of an injustice; along the way we regain and maintain our self-respect (we are not considered a coward - for a while at least); we have the moral high ground.

A *prolonged desire* for vengeance, thus, can give us more 'pleasure', and perhaps even emotional-cognitive direction and engagement, through morality fantasies, heroic daydreaming, strategic planning and brotherhood socializing than the actual taking of vengeance could ever do. In fact, the taking of actual vengeance would not only put us in physical danger but also *deprive us* of the above, and of our anticipated martyrdom for showing 'moral' restraint, and of our 'right' to make claims for legitimate justice. Indeed, it is our seeming moral restraint, linked with our resolve to see justice done, which attracts sufficient support for us to carry on in this way – we are not violent and vicious; we are seekers after a higher truth. So, in effect, the opposition is faced with such an increasingly massive physical and 'moral' alliance that vengeance may not even be necessary; demands for compensation become not ignorable; deference, on their part, not unthinkable.

Another possibility is that a powerful patron may hear of our plight and its general support, and in order gain favour and/or to prevent trouble (we never give up making measured threats and claims for justice) suggest taking vengeance for us, or simply threaten to do so, or set out to collect compensation for us, or let it be known that feuding will not be tolerated on his patch, and that justice shall be done. Whatever the case, we come out moral winners, we are safe and we can more or less safely continue to gossip about the injustice perpetrated by our adversary – the bully - and how we saw to it that he was sorted out. We may, of course, have to be nice to our vengeance allies for some time to come, or we may have to subordinate to a powerful patron well into the distant future. But that may not be a bad thing. We now have committed friends for future protection; or we have come under the protection of the powerful patron - we are no longer alone, we are linked to power. We have government, and it has ruled in our favour.

In as much as this scenario was driven by a 'restrained desire for vengeance' in which a lot of the pleasure was derived from thinking about vengeance, rather than from actual physical vengeance, there is, of course, the possibility of becoming strongly addicted to fantasies about taking vengeance, to the point of not being able to do much else but to talk of 'revenge and failed-justice', or of being crippled by obsessive hate and moral righteousness. This may especially be the case if public support is not forthcoming. Nevertheless, even prolonged, possibly addictive, fantasies of taking vengeance, usually relieve us of the burden of mentally subordinating to the initial perpetrator of harm against us, and our resultant bravado and constant threats of future vengeance may well save us from being picked on.

Whatever the case, during fantasies of taking vengeance we spend a considerable amount of time evaluating our enemies, possible allies, and, indeed, the safety and dangerousness of our proposed and past actions. During our dreams, daydreams and contemplation, we project our own actions into the future, anticipating both pleasure but also possible dangers in presumed reaction to them. As noted, our capacity for guilt and empathy may even lead us to consider the *fairness* of our *own* actual or proposed vengeance actions. Thus, our desire for vengeance, instead of motivating uncontrolled brutality, can lead to philosophizing about notions such as good and evil, morality and immorality and spiritual and political solutions to the human condition. Desired behaviours based on these conceptualizations can be postulated as hypothetical behaviour, and these projected into the future. And the grey areas within these, and the hierarchical gradations concerning types of behaviour and justice and morality will come under scrutiny. We may, for example, consider the possibility of an eye for an eye instead of total destruction of the other or measured as opposed to unlimited compensation; we might begin to theorize a political arena where vengeance itself is replaced with requiring individuals to answer before a higher code of justice based on *degrees* of fairness; one in which we are able to satisfy our desire for vengeance by seeing criminals get their due at the hands of the *law*, governmentally enforced. This, however, clearly implies the acceptance of *governmental authority*. Whatever the case, in so far as strong emotional-cognitive processes such as anger, love, hate, envy, jealousy, shame and depression are involved in contemplating vengeance and its ramifications, our eventual notions of the rights and wrongs of various behaviours and judicial solutions to conflicts are often highly charged emotionally.

Another scenario, from the start, would have been to 'show the other cheek'; to forgive one's enemies in the belief that one day such forgiveness would spread and result in a perfect world. Given what we know so far about human nature and human history, however, this tends to let cheaters, bullies, hypocrites and liars run rampant; it makes non-bullies forever victims. So, the *desire* for vengeance has not, despite all the complications discussed above, diminished during human phylogenetic history. Its manifestations, however, have been greatly tempered, altered and focused during the evolution of human nature and human social relationships. The result is a number of crucial aspects of political society, aspects which can provide a significant degree of security and social order. At base these key dimensions of political society are derived from a relatively simple desire for vengeance; but as we have seen, this desire becomes rationalized into public opinion, morality, right and wrong, laws and governmentally administrated rules and procedures. And, as a corollary, one of the major human nature forces for the willing surrender of authority (and individual resources) to governments to enforce judgments, has been identified.

But human nature has also evolved a variety of other means for roughly maintaining social conditions so that conflicting emotional-cognitive motivators and conflicting desires and fears (and conflicts within self-images) do not become so threatening (or inviting) in their expression that a bulling or vengeance scenario becomes a likely result. A major means of doing this, of protecting self in social situations, is through a number of emotionally felt mental restrictions. These motivate self-imposed injunctions, personal rules/taboos, social taboos, laws and social boundaries (in the first instance, self-control and self-identity) which motivate and guide individual behaviour. These too provide the basis of a great deal of larger scale human social behaviour.

CONTROLLING SELF AND OTHERS

Humans generate *self-restrictions*, then, as a first defence against those excessive desires and fears which might get them harmed, rejected or otherwise burdened. Besides generating personal taboos, rules and boundaries, humans attempt to get others to accept self by 'agreeing' to a number of *social* taboos, rules (laws) and boundaries. These give self a notion of 'who to approach' and a degree of reliability in the form of guidelines, guarantees and protection in the larger social world. Taboos, rules and social boundaries are, to some extent, based on expectations. We know that certain things are best *not done*, or at least *avoided*, if our expectations are to have a chance of being fulfiled. We do set goals and then attempt to concoct procedures and to design more or less self-contained (bounded) groups or organizations which we hope will fulfil them.

To a considerable extent, however, 'real' taboos, rules and social boundaries are derived from relatively strong bio-electrically developed attractions and repulsions. These processes are greatly influenced by pre-knowledge and are processes in which the distinction between conscious and unconscious processing is very often far from clear. Various mixtures of, for example, lust, love, jealousy, anger and hate - along with the 'right blend' of, say, fears of rejection, engulfment and humiliation - intermixed with desires, such as to be included, to gain and/or maintain status and to be respected – often result in highly charged cognitively derived images of, and procedures for dealing with, other human beings. Besides procedures for dealing with others, they include delineating both physical and psychological space between self and others and attributing to others more or less safe and dangerous characteristics. These images, procedures, spaces and boundaries can be sufficiently strong so that powerful feelings of security, or conversely, anxiety and danger, arise when they are encountered. We feel, in our gut, that we are safe because certain procedures (sometimes rituals) and categories (sometimes symbolic representations) will protect us. In other cases, we know, equally viscerally, that it is dangerous to proceed in a particular way or to approach a particular group.

Individuals, types of individuals and/ or groups become identified with various *charged* procedures, attributes, categories of humanness, rules and boundaries. Specific identifications represent, to a large extent, the assumed power of various individuals to deal with the potential dangers contained in human social experiences and their supposed capacity to fulfil desires. As a result, this process designates the general physical, social, political and spiritual desirability, or danger, of individual humans vis-à-vis oth-ers, including self. The results are the emotionally charged cognitive evalu-ations of objects, behaviours and humans discussed in Chapter Six. The con-sequences of these are species typical stereotyping, species typical notions of human sexuality, of common patterns of deception and self-deception.

Additional results of this process include patterns of hierarchy in human social relations, whereby humans are not so much separated and distanced by virtue of the procedures 'required' when dealing with them or by being included or excluded from groups, but by virtue of a charged network of hierarchical relationships in which moving about too much can be emotionally precarious. The result is a degree of perceptible social stability and predictability. From interacting within these networks individuals develop their personal version of what is to be desired and feared (within the human range) and their own *expectations*, often felt to be *rights*. When hierarchical evaluations are generally shared they become more than individual perceptions but also 'social facts'. They become generally agreed social distances, taboos, rules, duties, obligations, demands and boundaries. This gives a degree of predictability (and some control over) individual life circumstances. As in vengeance seeking turned into governmental action, conflicts concerning all the above can lead to formalization taking place so that human nature propensities become laws and officially enforced procedures, behaviours and rights. Hierarchy usually ends up in politics.

Human political society, then, is made up of the dynamics of taboo creation, rule making, the formation of boundaries, balances of power and hierarchy formation. And the more complex societies become (more people living in high densities, more complicated tools/toys, and so on), the more individuals have to run hard to stay even, and the more *political* social life becomes. Politics is involved in judging, enforcing, adjusting, creating, defending and discriminating among the various emotional-cognitive and desire and fear processes which pour out of the not altogether synchronic human nature described earlier. This is the basis of the argument that politics is the guiding principle behind the species specific processes which make up human societies.

Taboos and rules (charged distancing and procedures) and boundaries (charged category designations), hierarchy (charged distancing) and politics, then, are means for identifying, defining and reacting to (often preventing ourselves from entering into) circumstances which evoke feelings of danger. They, of course, also provide means of coping with dangers and for attempting to fulfil desires - while, inadvertently, creating the basis for human social order. Let us consider these processes in more detail. Taboos and rules will be considered in the rest of this chapter, boundaries, hierarchy and politics in the next.

Taboos

"The English word taboo is derived from the Polynesian word 'tabu'" where it generally meant 'to forbid' or 'forbidden' (Radcliffe-Brown, 1952b, p., 133). According to Franz Steiner (1967), in his classic study of taboos, "One might say that taboo deals with the sociology of danger itself, for it is. . .concerned with the protection of individuals who are in danger. . ." (pp, 20-21). Social taboos, thus, identify individuals, objects and behaviours which are to be avoided if safety is to prevail,

> ". . . For until taboos are involved, a danger is not defined and cannot be coped with by institutional behaviour . . . Taboo gives notice that danger lies not in the whole situation, but only in specific actions concerning it".

(Steiner, 1967, pp., 146-147).

So, it is not sex which is dangerous, but certain kinds of sexual presentations or specific sex acts; it is not conflict but specified kinds of hurting; not thinking but certain thoughts; not reading but certain books; it is not desire but uncontrolled desire (or certain kinds of desire); not humans as such but sinners, and so on, which are dangerous (see also., Thody, 1997). Sometimes desire, behaviour and thought are combined.

> "Ye have heard that it was said. . .Thou shalt not commit Adultery: But I say unto you, That whosoever looketh on a women to lust after her hath committed adultery with her already in his heart" (*Mathew* 5 - Authorized King James).

Taboos also act as mechanisms for generating obedience and for restricting behaviours in dangerous situations and thus result in ". . . the protection of society from those endangered - therefore dangerous - persons. . ." (Steiner, 1967, 21, 147; see also Davies, 1982). If we change this to read something like, 'protection from each other', we will avoid the teleology of 'society needing protection' and still be able to use its insight. Humans often find themselves in precarious situations, especially precarious social situations. It is here that potential psychological (and some would argue, social) dangers are ever present; and it is here that taboos are common (Radcliffe-Brown, 1952b; Douglas, 1970).

In effect what is being argued is that, for humans to avoid danger it is often necessary to avoid the danger of being a danger. Taboos and rules assist this process by causing us to stay away from certain situations and/or to follow specified procedures. Those who violate taboos are, thus, often thought to become of less value as a human for having done so (Radcliffe-Brown, 1952b; Douglas, 1970; Thody, 1997; Twitchell, 1987). Various forms of purification (ritual – and real - subordination) may be necessary in order to reintegrate such individuals back into social life . However, taboo violations may be so polluting, or corrupting, that the only 'solution' for others is avoidance through a more or less permanent separation from the violator (or violators). In the longer term, humans often develop specific notions of avoidance of certain others, behaviours and events so that the opportunities for danger are reduced on a more permanent basis.

Avoidance

Thus, taboos and rules designate a number of required and restricted behaviours which tend to keep those who internalize them out of danger and/or prevent them from becoming dangerous to others. This process, however, is more than simply a rational or practical activity; it only works effectively when, and if, certain objects, rules and regulations become charged spaces, arising from feelings of danger - fear of rejection, jealousy, guilt, envy and the fear of envy, for example. So, for instance, in societies in which threatening personal and social conflict might be expected between a mother and her son-in-law over demands for emotional loyalty from a female who is both daughter and wife, taboos have evolved so that a man, when possible, avoids his mother-in-law (Radcliffe-Brown, 1952a). In modern societies degrees of intimacy and, conversely, avoidance, with in-laws are established, and maintained, through the differential use of terms of address such as mom/dad, Mary/Bill and Mrs./Mr. Smith (not to mention nicknames). This use of differential terms of address applies not only to interactions with in-laws but also with parents, friends and enemies, not to mention between bosses/workers, patrons/clients, teachers/pupils, and so on, as means for intimacy, status recognition, patronage, stigmatization and for creating and maintaining social distance. It is very difficult to use any term other than the one an individual feels comfortable with; an emotional, viscerally felt, barrier has to be overcome to use terms very different from those which have evolved with given relationships.

In the vast majority of human societies it is taboo, except between adults who have a socially accepted sexual relationship, to observe people in certain states of undress and/or states of sexual exposure (Symons, 1979; Thody, 1997). This taboo results in the avoidance of all sorts of potential feelings of envy, jealousy, embarrassment, shame, inadequacy, unsettling temptations, guilt and undesired social conflicts. And, for individuals who have acceptable sexual relationships it is taboo to enjoy them in public – probably in all societies - again resulting in the avoidance of all sorts of jealousies, envies, comparisons, humiliations and embarrassments. Some sexual practices are themselves considered taboo (Cf., Thody, 1997; Davies, 1982; Douglas, 1970; Twitchell, 1987), and as such are done only in secret, if engaged in at all. A number of terms which designate particular sexual practices are established as vulgar/rude words, and in public generally and in the company of those who would be insulted by them, or might be embarrassed by their connotations, the emotional-cognitive restrictive power against using them can be very strong. Terms which designate certain sexual practices taboo are often generalized to describe disliked individuals (see below). When people rebel against such restrictions, or wish to glorify a stigmatized identity, on the other hand, they go out of their way to use these words as often and as emphatically as they can.

Humans do not usually help themselves, un-invited, to their friend's food, clothing, favourite war club or toothbrush. Individuals are given space, their own sleeping mat, prayer corner, room, office, car, telephone, computer, kitchen, doll, drawer, beer mug; and it is very difficult for others to violate that space and/or the boundaries which set it apart from the rest of the world. During attempts to do so violators feel uneasy - in the pit of their stomachs there is fear, a fear born not of rational knowledge based on an abstract concept of private property or individual rights, but fear generated from the charged spaces, taboos and boundaries that have been created between self and the personal effects of others. We know that we would be justifiably subject to the wrath of the other if found out. This represents a species typical emotional recognition of the precariousness of individual autonomy; 'there but by the grace of God go I', we feel when we see others deprived of control of even their smallest personal belongings. So, we emotionally charge spaces between self and others as protection against the powerlessness we, ourselves, so often fear when we interact with others, especially relative strangers..

Indeed, humans often feel it prudent to avoid certain types and/or groups of people altogether. These may be members of specific religious, racial, ethnic, status and/or political groupings. Aristocrats generally look to avoid the polluting effects of the lower orders, religious and political believers want to avoid the corruption of non-believers, ethnic groups are wary of the immorality of other ethnic groups, racial purists are apprehensive of the aphrodisiac powers of individuals of other races, serfs fear the aura of aristocrats, parishioners the awesome spiritual (presumed moral) power of the priest. There is often a very large symbolic/sign content in these types of avoidance. Dietary taboos, for example, are common.

> "Better, however hungry you may be feeling, to eat something which you know is in keeping with the customs of your tribe, than yield to blind appetite and forfeit your right to membership. . .[moreover]. . .By refusing to eat foods which they too willingly consume, we demonstrate how superior we are to them through the fastidious nature of our tastes." (Thody, 1997, pp., 95 and 10; see also, Douglas, 1970, 1975; Davies, 1982).

Young Alexander Portnoy, in Philip Roth's novel, *Portnoy's Complaint* (1971), was able to observe that this was not just the case with royal families of old, the ancient Hebrews or high caste Hindus, for example, but that it is still exists, at least in his family, in modern America.

> "There isn't enough to eat in this world, they have to eat up the *deer* as well! They will eat *anything*, anything they can get their big *goy* hands on! And the terrifying corollary, *they will do anything as well.* Deer eat what deer eat, and Jews eat what Jews eat, but not these *goyim.* Crawling animals, wallowing animals, leaping and angelic animals – it makes no difference to them – what they want they take, and to hell with the other thing's feelings (let alone kindness and compassion). . . [they] know absolutely nothing about human boundaries and limits (p., 90-91).

With regard to deer, the above attitude is not, in fact, unique to Portnoy's family and acquaintances. In England deer are fed to the hounds not consumed by the 'gentlemen' and 'ladies' who hunt them, while in the 1978 American film, *The Deer-Hunter*, it was ". . . clear [that deer-hunting] tends to be a lower-class activity" and that the killed deer are eaten by such people (Thody, 1997, p., 111).

In certain circumstances throughout human history all of the above mentioned social separations and supporting taboos have been common; various boundaries based on them have at one time or another been seen by historians, anthropologists and sociologists (as well as participants) as being basic to human 'social organization'. The purpose here, however, is not to identify the circumstances in which any of the above is likely (as important to a social science as that will be), or to make value (moral) judgments concerning various separations, avoidances or boundaries. Rather it is to argue that all the 'normal' and common human interactions discussed above are conditions in which taboos have abounded (and continue to abound) and that, therefore, the processes which underpin the taboo process are basic to human social life. They are, moreover, basic as a force emanating from human nature, not from culture or some mystical notion of 'the needs of society'. Indeed, the picture of human nature which emerged from Chapters Four, Five and Six strongly suggests that the characteristics of taboos and avoidance discussed above would be widespread in human social life. In so far as taboos and/or patterns of avoidance are related to 'social organizations', in an ultimate sense they are *causes* of them rather than their products. Of course, taboos and 'organizations' become part of a feedback relationship more or less from the start.

It is worth noting that symbols and signs such as words, appearances and foods, especially lend themselves to the taboo process because they can have an immediate impact on the human senses. They are heard, seen, felt and smelled. Therefore, 'positive' and 'negative' bio-electric charging is relatively easy during the learning process. The same can be said for certain body expulsions such as snot, excreta, body odour and menstrual blood, for example, from which it is very easy to develop a notion of dirty/polluted to apply to individuals or to specific groups (racial, ethnic, status, occupational) to justify avoidance and separation. Sometimes certain activities generate visual, olfactory and symbolic impact all at the same time. Negative attitudes towards cigarette smoking, for example, are not exactly new. Henry Ford argued that "If you will study the history of almost any criminal you will find that he is an inveterate cigarette smoker." A similar view was held by John L. Sullivan, famous ex-boxer, "It's the Dutchmen, Italians, Russians, Turks and Egyptians who smoke cigarettes and they're no good anyhow" (Tate, C., 1989, pp., 111 – 112). But tabooizing and stigmatizing are not the only relatively subtle, often symbolic, means humans have of avoiding dangers in social life.

Joking

Where physical and social avoidance of potential danger spots is not feasible in interpersonal relations, joking relationships are often established so that encounters are divested of their volatile potential before blood is shed (Cf., Radcliffe-Brown, 1952a). This is as true of modern societies as it was of the ones studied by Radcliffe-Brown. I doubt that any known human relationships, from sexual partnerships to international diplomacy, would be possible without a considerable amount of joking. Above all, we feel we can trust humans who have a sense of humour and can laugh at themselves; we find it difficult to feel comfortable with people without a sense of humour. This is because a sense of humour requires an understanding of at least some of the dilemmas, paradoxes, absurdities and conflicts involved in being hu-man. Moreover, being able to laugh at oneself demonstrates that one is ac-cepting that self is not perfect, that self is capable of understanding and forgiving failure. As such, humour plays a major role in preventing a numb-er of head-on collisions and thus is a significant means of avoiding social dangers and cementing relationships.

A sense of humour, moreover, gives a sense of control (Cf., Girard (1978; Freud, 1957; Davies, 1990) to individuals who might otherwise feel extremely vulnerable. This is clearly illustrated in Beaumarchais' play, *The Barber of Seville,* when Figaro says, "I force myself to laugh at everything, for fear of being compelled to weep". During the practices of 'creative' sub-ordination and/or domination joking can play a major part. In social hierarc-hies, generally, there are often a number of points of potential friction (and thus danger for the participants) between formal power and status (authority and expected social deference) and hoped for co-operation (requiring a de-gree of intimacy) which are eased with joking – such as between lecturers and students, doctors and patients, priests and parishioners, for example. Through joking a specific formal relationship is maintained, but even the subordinates can feel a sense of control of encounters, or at least of being welcomed to participate in them. Moreover, both sides in the above exam-ples can joke about the others in private and thus feel a sense of equality, if not superiority, to them.

For example, despite years of teaching the 'rational' principles of 'socialist realism' and 'scientific socialism', along with the virtues of co-operation and the morality of sacrificing for a non-exploitative state, it was said that a common phrase in factories across eastern Europe during the Communist period was, 'they pretend to pay us, we pretend to work'. This bit of wry humour conveys considerable information; it also provides a basis of solidarity for workers to shirk and not feel obliged to tackle the establishment. In other words, it allows for the *avoidance* of potential life threatening conflict without a loss of self-respect. Jokes, can in some circumstances, question governmental policy in ways which can have considerably more effect than direct political challenges. For example, a joke about the old Soviet Union which circulated in the west during its final years went like this: a man in the Soviet Union was buying a car; after the deal was finalized he asked the salesman when the car would be delivered. He was given a date four years, three months and seven days hence. 'Will it be in the morning or afternoon?' asked the buyer without hesitating; taken aback, the salesman asked, 'Why, does it matter!?'. 'The plumber is coming in the morning', answered the buyer'. This speaks volumes about central bureaucratic planning and postulated outcomes in terms of service. It says much which is easily taken in and absorbed by individuals who would not wish to study complex production, distribution and resource allocation statistics in order to argue that central planning and distribution on such a scale as the Soviet Union attempted may have its problems.

And when a president of the USA during one of its most non-imperial presidential periods, famous for his few words and seeming lack of action (Calvin Coolidge), died, a news reporter asked the person reporting the news, 'how can you tell?', the answer was, 'his cheeks are rosier than normal'. By all accounts Soviet economic planners and Calvin Coolidge were not stupid people; nor were they especially hated – at least by those who made jokes about them. But there were those who wished to challenge (through ridicule and by influencing the public) their approach to government policy without themselves becoming directly involved in the political process. In the case of the US president, this joke represented a clear view that complete *laissez faire* economics and market operations were, maybe, not the best way forward and, more specifically, a view that a president of the US should be a dynamic leader not an absentee chair of the board.

Humour is a very democratic means of making such points in that it reaches many parts other methods do not. Above all, it allows common people an opportunity to be anti-government without the personal dangers of struggling to write a grand political treatise, working endlessly for a party, joining a revolution or dedicating their life to attaining high office. Thus, however useful it can be, any information gained by politicians from humour is a by-product, rather than a cause, of political humour. Humour for humans is those processes – largely involuntary and subconscious - which represent reactions against danger. Jokes include complaining but in such a way that it seems extremely small minded to retaliate with physical, social or legal retribution. This can have significant implications for human social life. In politics, for example, when the top people are no longer subject to jokes, democracy and free thought are in trouble (Sweeney, 1997, p., 24);

when it is impossible to joke about leaders and governmental policies the alternatives may be clandestine subversion of the implementation of policies, secret societies, anti-government conspiracies and riots; not to speak of a loss of faith, trust and respect for a particular approach to government itself. To appear safe, to appear as: government that is a worthy patron; government that is of the people, for the people, government too, must have a sense of humour, that is, show some attributes of humanness.

Jokes directed towards governments, then, can be a barometer for individuals in positions of authority to gauge the success (or lack of it) of particular programmes, organizations and procedures, allowing leaders an opportunity to avoid unpopular policies. To cut out this source of information is not always wise from the point of view of leaders, however hurtful some jokes might be. There have been circumstances in human history when any opposition to specific governments, let alone joking about them, has been seen as being intolerable, on *rational*/ideological purity grounds. But, like trying to get rid of emotions, the sex urge, dreaming, fantasizing, thinking and drinking, trying to eliminate humour is usually not successful.

Moreover, it can be argued that the process of humour is extremely important in establishing avoidances, taboos, rules and even social boundaries. Parents tease children in order to establish restrictions in a less hostile and more loving way than direct injunctions might do. Teasing uses statements which, depending on how they are said, can be encouragements, signals to stop and/or suggestions for altered behaviour; to try to make children more or less outgoing, more or less clean and well groomed and more or less active/competitive, for example. Adults call each other to heel, and so establish claims to rights, with slight teasing and/or the use of running jokes. Bosses discipline workers and workers tell bosses that they are out of line, friends remain friends while renegotiating relationships through humour.

So, humour has also been used in human social life to do more than avoid danger and as a means of political communication; it has also been employed to enforce taboos and social boundaries. Within this process there has, moreover, been a tendency for humour to suggest superiority and inferiority of individuals, groups and, indeed, nationalities. For example, it was reported in news magazines during the 1970s that Germans (East) spoke of a smoke filled room as being full of 'Polish air', or a total economic disaster as being 'Polish economics'; and it was rumoured in America that the first ever Polish pope's initial miracle was to turn a blind man lame (Davies, 1990a). This is clearly saying something about Poles. Now, given that, historically, national Poland (not an ethnic unity by any means) has never achieved much in the way of long term nation building (or independence), that it is surrounded by powerful, often warring, neighbours who prefer to fight in Poland rather than in their own territories, that much of its land topography is good tank country, and given that very large numbers of Poles have lived as a marginal ethnic group in the US, it is, perhaps, no wonder that Poles are the brunt of so many ethnic/national jokes.

But, any comparative look (Cf., Davies, 1990, 1990a) quickly dispels any notion that Poles are in any way unique in this regard. Also, it certainly is not the case that nationality or ethnicity are the only dimensions of human existence which are the basis of jokes which set out types or groups of people. Although very common and significant, nationality and ethnicity (Cf., Davies, 1990) - Poles in America get hit twice here – are joined by race, professions, locations, gender, certain sexual practices and/or sexual orientations, for example, in this regard. In all cases, social taboos, separations, avoidances, social distances, boundaries and 'moral lessons' are not only reinforced but passed from one generation to the next.

"*Mountain Mother*: My, Billy. Your prick is bigger than Dad's!
Billy: Yes, that's what sister always says." (Twitchell, 1987, p., 53)

This joke not only stereotypes and stigmatizes individuals from a rural backwater, but also reinforces, through ridicule, the powerful taboo (common in human societies - Cf., Fox, 1967, 1980; Thody, 1997; Twitchell, 1987), against incest; at the same time it lends support to taboos against other forms of anti-social behaviour. Twitchell points out that by putting the joke in a context in which few people have experience, an out-of-the-way place, it is truly a joke; one which is able to "pass censure together with forgiveness, or at least with understanding" (p., 54; he is following the argument of Legman, 1971). It is a joke; it only happens among hillbillies – who are somewhat beyond our comprehension anyway; in our own context, in our real life, we would condemn the above youth with "The most obscene and ferocious curse in the English language. . . "mother-fucker"" (Twitchell, 1987, p., 54), and then we would apply the *term* to individuals for a variety of non-sexual, but detested, activities. Here, then, we have an example of a joke which co-exists with, and supports, one of the most powerful taboo words we have. Both are based on powerful feelings of sexual repulsion which we use to stigmatize, if not prevent, more than just mother-son incest but a whole range of behaviours and individuals who are deemed generally beyond the pale/comprehension.

Humour, then, is a fundamental ingredient in most social situations. It is a hint of what is in the subconscious mind, as Freud (1905) so astutely observed; but it is more; it is part of the way humans live life without killing themselves or others in the process; it is part of creating and reinforcing taboos, rules and social boundaries. Laughter itself is part of the process of coming to understand, and accept, the human condition: "I would go so far as to venture an order of rank among philosophers according to the rank of their laughter – rising to those capable of *golden* laughter. . . ", proclaimed

Nietzsche (1990, p., 218, his emphasis). He continued with the observation that, indeed, Gods themselves laugh at all serious things when they philosophize the human condition: "Gods are fond of mockery: it seems they cannot refrain from laughter even when sacraments are in progress" (p., 218). The study of humour and joking (not just jokes) - and laughing - deserves a much larger place in the social sciences than it so far has been given.

Self-Accusation

Another method of avoiding danger is through the practice of making self-accusations in order to take the sting out of anticipated failure. 'I'm hopeless at tennis, but if you really want me to, I will play with you'; 'I know this essay is not very good but I can't think anymore. Here it is'. In these examples self-accusation carries with it a sort of counter-accusation (if I lose it is your fault for making me play tennis, and you are a bastard for wanting to humiliate me!; why is the essay question so confusing and unreasonable?). The second response, of course, borders on a bit of begging, a throwing of self on the mercy of the court, 'please sir, I tried my best'. This can grow into, 'I don't know what spirits possessed me'; 'I was temporarily insane',. In these there is a sort of, 'I know I messed-up but it was a blip, a one-off, it won't happen again; it is really not *me*, I am ever so sorry'. This shows a degree of humility which humans generally respond to in a positive way. If indulged in too often, and increasingly confirmed by others, however, self-accusations, especially of the variety of the last two examples, can lead to rejection and isolation. This can force a turning inward, which may lead to a self-definition of uselessness, motivating psychological work to strengthen the barricades around self. The result, nevertheless, is usually further restrictions on individual, externally oriented, 'deviant' behaviour, in order to avoid potential future dangers. Such self-imposed restrictions can have an external manifestation in that they may motivate, at least some, individuals into a strong adherence to the rules, taboos and social boundaries which seem to eliminate, or at least control, possible failure, loss and/or ambiguity.

Humans are, in fact, quite directly, often consciously, motivated in their adherence to rules, taboos and social boundaries when inherent dangers, or potential dangers, can be relatively easily identified and thus avoided. Humans, other things being equal, simply try to avoid participating in - or they try to prevent occurring - circumstances in which they have to put themselves on the line for judgment, risk suffering losses of respect, esteem, status and power (control) and/or have little idea as to what is required of them in order to maintain these. Such circumstances can thus be considered *prime social danger areas* for humans *in all cultures*. But, avoiding all of these circumstances of potential danger may become so restrictive of behaviour that humans lose out in reproductive and survival competitions. 'Nothing ventured, nothing gained', as the saying goes. And, anyway, we know from common observation that for many humans a propensity to jump into just such conditions, at least on certain occasions, can be quite strong. The evolutionary advantage would seem to be able to avoid circumstances when the danger is most likely or prominent, and to jump in when it is minimal.

Humans have, in effect, evolved an extremely subtle means of restricting behaviour in order to avoid many of these dangers, until, that is, their capacity to succeed has greatly improved, or at least such 'opportunities' have been evaluated and the potential dangers more fully assessed. This is the human capacity through fantasy to defer (sometimes indefinitely) gratification. As a result, traditional taboos, rules and boundaries are more or less adhered to, but not always - not when a realistic chance of fulfiling desires and avoiding fears comes along.

Deferred Gratification

It is tempting to argue that it is the human nature capacity for fantasizing self into a number of positively felt emotional-cognitive conditions in a future 'utopian' state which gives humans one of their most distinctive characteristics. This capacity makes immediate deprivation easier to take and more instant gratification deferrable. It not only avoids the dangers of attempting certain social activities without adequate evaluation but also the effects of certain reproductively disadvantageous addictions. It makes generalized altruism common (through sub-ordination to others, rules, taboos, norms and boundaries) as a means to gain aid on the way to a future 'good life'. Indeed, it not only makes these things bearable but, given a number of human desires and fears, often even enjoyable. It makes reciprocal negotiations likely; in other words, it makes humans social and political animals.

It is not that a future state of affairs will objectively provide gratification that is the motivating factor. It is the fact that *creating* and *daydreaming* about *fantasies* with self in some future state of happiness can be extremely rewarding in the here-and-now; more so than either the present, objectively assessed, or future existence ever turns out to be. Indeed, it is through this process that otherwise boring and undesirable tasks can themselves sometimes be rewarding because they become part of the fantasized road to the fantasized future state. In fact, suffering deprivation in early life almost glorifies the struggle for the good life; and in later life too, especially retrospectively if the individual does more or less succeed.

The human ability to receive gratification from daydreaming self into a future in which an individual interacts with beautiful people, warm companions, perfect sexual partners, cultural heroes and, even, mythical heroes, as a means of controlling immediate desires and envious feelings (and thus for often being able to avoid social dangers) has few equals. This is not unrelated to human nature propensities for feelings of impending 'Luck' (optimism) among those who feel that most things are in the hands of 'Fate'. Luck is felt to be a supernatural force which gives *quality* and *success* to the *future.* Its anticipation makes the here and now bearable (indeed, even very exciting for gamblers). For some, it is a sense that whatever is going on now, 'Fate' has (must surely have or 'why would I have been put on earth? – and moreover, why would I be so restrained?') in store, for me.

In addition to these motivators for deferring gratification, we can consider Carroll's (1985) argument that feelings of guilt are its basic motivators. He argues that the guilt developed during childhood from a fear of completely losing mother's love when being separated, motivates children to restrain their demands for total and immediate gratification. Such guilt becomes embedded in self. There undoubtedly is validity in this view. Whatever the order of priority of these causes of deferring gratification, motivation for concentrating hard in the here-and-now, and for putting up with a number of difficulties in the hope of achieving realization of a much better life later on, is provided. Mixed together these motivators very likely lead us to identify with individuals who seem to have achieved success in life. They probably also provide the major reasons for humans admiring fame and fortune, and even for worshipping cultural heroes. We want our future (fantasized though it might be) to be full of the types of people and places that legends, myths, advertising, television, films, travel books and sports display to us, not what we now have in the form of our competitors or our dull neighbours; we do not see the future as containing our existing worries and anxieties, our current mundane tasks to get through, or our drab existences.

The human ability for deferred gratification, then, may be extremely basic to human reproductive and survival success. It helps avoid early addictions and gives humans very long periods of time to dream and daydream about their future prospects before they jump into potentially threatening conditions. (It also generates creativity and myth making through fantasies and utopian dreaming.) At the same time, human sociability through its encouragement for the acceptance of rules, taboos, boundaries and hierarchy, is enhanced. However, this kind of fantasy development could itself easily have become an evolutionary trap in that humans could have come to spend all of their time dreaming and fantasizing rather than surviving and reproducing. Thus, we can speculate, there was an evolution of a number of addictions, desires and fears which could only be satisfied with more immediate gratification. We have, as evidence of this, for example, a powerful capacity for lust, love, feeling envious, hatred and jealousy - built in-to a number of more complex fantasy/desires and attendant fears - all of which play havoc with excessive gratification deferring and utopian dreaming.

Deferred gratification based on fantasy can also lead to a sense of non-fulfilment in that, because fantasies can never be completely, if at all, fulfiled, a day of reckoning is likely to come. At some point individuals will begin to wonder if they have not been fooling themselves, have they been conned by others, they wonder. Usually, at some time or other, a sense of deprivation is felt, rights seem to be violated, expectations not fulfiled. (Formal utopian ideologies, in fact, continually present the present as being an undesirable state of affairs [Mannheim, 1960], often pictured as being full of exploitation and injustice designed to keep people subjugated.) It is for all these reasons, we can speculate, that humans have evolved a surprisingly good ability to be realistic in formulating their expectations.

Expectations

If we define expectations as fantasy/desires which we think should come to fruition at some future date - as long as we prepare properly and defer our gratifications now - we can tie the concept of expectations to the notions of avoidance and social restraint we have been establishing above. Expectations differ from pure desires and/or fantasies in that they are based on observations of available objects and at least a degree of progressive achievement towards them. That is, excitement and heightened sensitivity during daydreaming about expectations - and during the deferring of gratification - have been progressively encouraged. Nor, we can hypothesize, has daydreaming been tempered by the development of too many fears of potential dangers to their fulfilment. Furthermore, it is likely that there has not been an excess of taboos or restrictive rules (charged spaces) encountered along the way.

Theoretically, expectations are our emotional-cognitive evaluations of our reproductive potential given both our species typical drives, emotional-cognitive processes, desires, fears, experiences of deferring gratification, specific opportunities and our encounters with taboos, rules and boundaries. Expectations come to designate areas where self is willing to be put on the line because self feels safe in doing so. Thus, we have a reason for the commonly made claims that human *rights* have been violated when expectations are blocked. Such claims are emotional-cognitive charged social statements which proclaim that self has been violated. This, from self's point of view, may well be the case. A lack of fulfilment, however, can also generate dangers within self. Claims of self having been violated may, in part, be derived from unconscious (or even conscious) fears that, in fact, more taboos were ignored than was safe during the generating of, and attempting to fulfil, expectations; claims of being violated may be a cover for fears that self may have made a big mistake in setting such expectations in the first place. To minimize these dangers humans often look to 'guidelines' in the forms of norms, ideas, values, a cultural ethos or ideology.

These concepts have been subject to a considerable amount of debate in the social sciences as to the role they might play as causes of behaviour. I do not wish to enter into the details of these debates but rather to suggest, firstly, that these do not, indeed cannot, exist as *sui generis* abstractions deriving from some mystical black box in our heads; and secondly, that they should not be treated as causes of human behaviour except in so far as they derive energy or force from the deeper feedback processes discussed in this and the previous two chapters. What is at issue is the relationship of such things as neural bio-electrics, emotional-cognitive processes, dreaming, daydreaming and the more abstract, symbolic abstractions called ideas, norms, values, beliefs and so on. In one sense clear distinctions between these processes are not possible. Ideas, for example, are biological facts. They are neural interconnections (knowledge units - images) generated from complex interactions of the bio-electric processes derived from neural excitation and inhibition as these generate mood states. However, their abstract and symbolic nature is usually quite consciously recognized, making them less permanent, more easily created and changed. They are much more a surface presentation of the phenomena discussed so far in this chapter than representations of anything real in their own right.

Our tendency to fix on specific types of lust and love objects, for example, is much stronger than our professed beliefs about these things. The first usually remain for life. They are tenuously repressed through very difficult experiences of love, envy, jealousy, guilt, hate, and so on. The second, such as, for example, a belief that sex should only take place within marriage, can change very easily - perhaps because of falling in love or getting a bit older. Many of our personal anxieties and fears concerning presentations of self and status never leave us, but an idea that we should behave in church or school can very easily come and go. The idea/belief that smoking is dangerous to health can catch on very easily even when the addiction is too strong to allow an individual to give up smoking.

Ideas, beliefs and values, like other knowledge units, when daydreamed and fantasized about, intermingling with desires and fears, and reflected upon, do become parts of self and are often expressed as personal expectations. They are felt to have meaning. But such meanings change relatively easily while an individual's basic ‘personality’ remains much more constant. In other words, an individual's fundamental emotional-cognitive propensities, desires and fears represent relatively strong bio-electric, often unconscious, feelings, while ideas, values and beliefs represent more conscious, analytical categorizations. As such, ideas, values and beliefs represent one manifestation of the link between deeper bio-electrics and everyday behaviour. Ideas, in this sense, can be important causal factors, not because they are symbolic abstractions with their own independent compelling motivational power, but because they are temporary, representations of, often highly charged, daydreams, fantasies, desires, expectations and feelings of what is safe or dangerous (good or evil, right or wrong, just or unjust); as such they are parts of a larger biological whole, not independent events. They are a manifestation of part of the natural kind of human nature.

Cultural ethos, then, has scientific significance as a vaguely related and socially shared set of historically and situationally influenced meanings about basic human emotional-cognitive processes, desires and fears. These meanings will include some symbolic development of ideas (knowledge unit elaboration) concerning expectations, senses of right, senses of justice and senses of what is safe and what is dangerous. Ideology has significance as a more coherently organized set of ideas, beliefs, values and prescriptions about the above, as shared by a more definable group with specific hopes, fears, or desires to paper over otherwise painful contradictions inherent in social life (Hamilton, 1987, 1989).

Even the most coherent of ideologies, supporting the clearest and most rigid of taboos and rules, however, is not sufficient to completely paper over all the cracks, to make humans feel perpetually safe. At best humans are only partially a taboo obeying and rule following species. Quite apart from inborn conflicting emotional-cognitive processes, environments change unpredictably. Envy and jealousy, for example - by self and by others - are likely to break out at any time; the control of lust, hate, envy and jealousy has to be on-going, and the taboos and rules established to aid in this task need constant vigilance. Desires for love, status and power can become excessive, fears of failure, exaggerated. Emotional-cognitive management is a never-ending process, and in the end no amount of avoidance, guilt and/or deferred gratification, will ever, by themselves, solve the problems of uncontrolled emotions, desires or fears and their potentially dangerous social manifestations.

Rules

Nevertheless, the attempt to establish clear taboos and rules is very often a *human ideal* in the task of emotional-cognitive management. Humans often attempt to get inhibitions, restraints and avoidances, for example, formally accepted as codes of practice or laws. The politics is in the direction of getting legal *enforcement* for taboos, habits, customs and so on. This can be argued from - and may make sense from - a rationalistic point of view, but in practice it is considerably more difficult and the 'rationality' less clear. If custom and informal taboos are not sufficient, it is likely that there is considerable disagreement among those theoretically subject to them. In other words, desiring taboos to become rules and laws is probably a sign that the taboos are breaking down and so the forming of laws is, from the start, going to be a tricky matter. Disagreements are most likely to be very emotionally felt. If laws are to apply to everyone (and few governments fancy the idea of enforcing laws which have to be differentiated on the basis of too many variables), the jurisprudence which goes into their formation can be quite complex, and politically divisive.

The point here, however, is not to enter into complex issues of jurisprudence, but rather to point out that laws and rules are just as much vulnerable to unsettled human nature as taboos or customs. The claim that they somehow represent an evolution of a greater degree of rationality is largely misguided. To many in pre-industrial times they probably seemed less 'rational' than taboos. After all, everyone 'knew' why taboos existed, it was much less clear as to why the Church or Baron wanted particular laws, except for their own self-aggrandizement, or something like that, it was undoubtedly suspected. In modern times clubs, cities, companies, militaries, professions and sports bodies, for example, lay down rules, codes of practice and regulations, very often to the effect that participants see them as obstacles to be overcome rather than as having the seemingly more untouchable force of taboo or the compelling imperative of a higher rationality.

Still, humans often create rules/laws and follow them. The underlying motivation is probably more in terms of the human propensity to classify and to organize things and to set up procedures for giving predictability to a variety of situations than a search for a higher rationality. And this is not necessarily a bad thing; clear procedures have the capacity to provide security and, in fact, to be 'fair' in so far as, once established, they can be made to more or less treat all people the same. Humans like to feel that the circumstances they enter into will remain predictable at least until they have left. In other words, rule generating is yet another dimension of the human will to power; policy formation, for example, is an opportunity for some in an organization to impose their will, for others to impose theirs by secretly flouting rules, and for still others, to find long term security. For those who feel that that have certain skills and attributes to show off, they like rules which mean that others can not easily cheat and so their own skills will shine through; most good footballers want to win by the rules rather than by cheating and 'shin kicking'. Those who fear that they have only selective skills and attributes like rules so that they know exactly what activities to avoid; and if they cannot do so, then at least they have the option of watching the other like a hawk and claiming, at the slightest opportunity, that the others did not play by the rules.

So, humans have a lot to gain by following rules. Human nature likes rules, especially ones which can be violated with deceit if self feels hard done by, but ones which can be imposed on others so that moral condemnation can be heaped upon them if they violate them. Rules and laws come from human nature; they are part of politics; rules and laws do not arise from the dictates of complex social structures, the evolution of rationality or some inevitable compulsion to progressive development. They do not represent a higher order of being; they are another way that human nature has of establishing some order on the surroundings of its major manifes-tation, self, so that reproductive behaviours can flourish. So, too, humans like social boundaries, and humans generate hierarchies, and inevitably be-come embroiled in politics, as they strive to generate predictability (and to some extent guarantees) in their quest for desire fulfilment and danger avoi-dance. These will be the subject of the next chapter.

SUMMARY AND CONCLUSIONS

In search of a science of humans we must link something like the concept of human nature developed in earlier chapters to a notion of 'political society'. This concept of political society asserts that human societies are the results of politics. Politics, as processes of negotiating, subordinating, dominating, (and a variety of other things discussed in this chapter), from self to the state, is where the conflicts of human nature are turned into social relations. This, of course, refers to both politics with a small p (politics of self-presentation, of self-worth, of sexuality, of child care, of interactions in the village square and internal military politics, for example), and also, state politics with a large P (party politics, factional conflicts for control or influence within the polis, state politics). Humans do not especially like to think of themselves as being emotionally motivated political creatures without a higher purpose; instead they like to think of themselves as being part of a greater design, or at least as being rational with the ability to construct efficient, perfected organizations and social systems (if not both). So humans follow a number of taboos which seem to suggest that they are following a higher spiritual calling, obeying the law of nature if not of the supernatural. But these are very much part of the human condition, derived from the conflicting emotional-cognitive reactions which drive humans, at the same time, to independence and, conversely, to subordination.

And when humans set out to be more rational with rules and laws, they are still not able to escape the reins of human nature and its inherent conflicts. Nor are they when they undertake to build boundaries. Yet, because all of these have served many reproductive interests, they are, in many ways, the human ideal. So there will be no stopping attempts to create boundaries; to enclose the 'real' and pure and exclude the deceitful and impure; to create structural parts of a larger, desirable edifice. But here too, human nature is the guiding force, as our considerations of boundaries, hierarchy, politicking and spirituality in the next chapter will show.

CHAPTER EIGHT
IN SEARCH OF POLITICAL SOCIETY

In Chapters Six and Seven it was argued that it is both possible and fruitful to classify human behaviour into 'natural patterns', or 'natural kinds'. Each pattern (or kind) will have to be based on 1) a relatively high degree of coherence based on identifiable emotional-cognitive motivators, desires and fears (human nature propensities), and 2) on their apparent anthropological and historical universality.

In Chapter Seven I began to consider a number of possibilities. For example, in seeking safety for self it seems that humans have evolved a propensity to avoid the reproductive drawbacks of both fight and flight. This was seen, for instance, in what seems to be a major human tendency when harmed or slighted to *delayed* vengeance seeking. Delayed the seeking of vengeance was seen to have a number of possible – but predictable – behavioural and political outcomes (patterns of behaviour), including the seeking of vengeance by governmental proxy. I also looked at human attempts to control and protect self via 'managing' emotions and/or excess desires and fears through capacities and propensities for tabooizing, avoiding, joking, self-accusing, deferring gratification, formulating expectations and a general mapping of the social world via ideas, beliefs and ideologies. Rule making was also identified as a powerful human nature capacity with strong supporting emotional-cognitive, desire and fear motivators.

As noted, in so far as combinations of these human nature propensities can be shown to result in identifiable, patterned outcomes, it can be suggested that they generate species typical natural kinds of behaviour – such as, for example, them-and-us groupings with internal rules of behaviour, village etiquette (customary law), patron/client relations, micro-hierarchies and governmental authority. To take our analysis to this next stage, the stage of analyzing how specific combinations of emotional-cognitive motivators, desires and fears (propensities/tendencies) turn into specific natural kinds of behaviour, is, unfortunately, not straightforward. Natural kinds deriving from these are rarely direct derivatives of them. Rather, they are most often products of interactions among several, not always harmonious, emotional-cognitive processes. Indeed, human nature motivators do not, as a matter of course, pull in the same direction. And, even when not in conflict, they can often be ambiguous and confusing. Understandably, all of this is reflected in behaviour, and, thus, in a certain instability of natural kinds of behaviour – some more so than others, of course.

Nevertheless - the argument here is - human nature propensities do generate sufficiently predictable human reactions to result in what can legitimately be considered natural kinds of behaviour (even though they are *more* or *less* durable and persistent depending on: the combined strength of the various emotional-cognitive processes which underpin them, their overall compatibility with each other and the environmental circumstances in which they exist). Natural kinds, in turn, are arranged, and rearranged (informed, influenced and limited by human nature propensities), to make up the ingredients of what can be called *political society* (at a given time). So, for example, we have seen how a desire for vengeance can result in a variety of behaviours, groupings and thoughts (but which are nevertheless limited, predictable and analyzable). Some of the results were alliances for blood feuding (arguably a 'weak' natural kind); in some circumstances, however, this desire can activate arguments for concepts of justice (very likely another human nature inclination based on feelings of envy, jealousy and fears of loss) and, in turn, for a legitimization of governmental authority (a natural kind – but one requiring more emotional-cognitive underpinning than just a tendency to seek vengeance and justice – see below). Groupings formed to take vengeance can (with the influence of the propensity to male bonding) take the form of war-party/terrorist-cult/political-insider groupings, which generically, it can be argued, represent a 'stronger' natural kind of behaviour than pure vengeance gangs.

Taboos not only identify danger in social situations and control individual behaviour, but give identity to groups, stigmatize outsiders, distance people from each other and generally 'civilize' people. All of this contributes to a number of natural kinds: from band living to church formation, to the establishment of mating patterns, kinship networks and nation states, for example. Motivations to set expectations and to defer gratification provide a basis for a realistic life style, if not a desired one. As such, both involve *evaluations* which help identify dangers in social life. They also go some way to establishing self-identity (arguably a natural kind as self/personality) which has some chance of confirmation - if not 'positive' recognition - even in the context of those dangers. But setting expectations and deferring gratification also provide a basis for agreeing with others what prestige, status, honour and privileges one can expect if one fulfils certain obligations and duties. So they are also a basis for accepting hierarchies, for politics and for wanting governmental authority (a natural kind) to judge, arbitrate and guarantee such *rights*. The human inclination to make and live by rules does much the same as taboos, with less emotional intensity perhaps, but rules are more easily changed to fit changing expectations; furthermore, rules lend themselves to the formation of more formal organizations and to hierarchies.

There are many more patterns which can be discovered in all of the above. But with the notion of rules, organizations and hierarchies we require an understanding of a number of additional human nature capacities and motivations (propensities), those which are perhaps somewhat more directly related to natural kinds of behaviour than the ones discussed so far. The first is the human nature capacity and inclination to build, maintain and be safe within the confines of social boundaries; the second is the 'package' of motivations which lead to the hierarchical *evaluations and the distancing* of objects, tasks, behaviours and individuals from each other; the third is the human nature attraction to the *games of politics* themselves (politicking) and the fourth is a human propensity to spirituality. A search for more natural kinds, as well as a more complete analysis of any of the above in terms of the working of political society, must incorporate an understanding of these propensities/tendencies.

SOCIAL BOUNDARIES

Social boundaries are, of course, based on taboos and rules and on emotional- cognitive charged social spaces between individuals and between groups of individuals; but they are also based on charged, cognitively derived images of distinct groupings, hierarchical layers and/or specifically identified boundaries. During emotional-cognitive development: 1) individuals and behaviours are more or less distanced from self and/or classified into stereotyped groups, often hierarchically arranged; 2) self develops the ability to 'feel it' when entering charged (tabooed) spaces thus giving potential boundaries recognition; 3) self also recognizes (feels) that a somewhat different presentation of self and/or behaviour will be required when living within boundaries. New taboos and rules will have to be followed and, as a result, self begins to accept, and impose, restrictions (defers gratification, for example) on its own behaviour; and 4) from this self begins to experience emotional-cognitive pressures for the establishment of specific criteria of inclusion and exclusion. A number of these points have been considered previously but in order to focus on social boundaries more directly let us consider each of these aspects of 'social bounding' in turn.

Social Classifying/Stereotyping

This first aspect has been relatively thoroughly considered in Chapter Six as it largely refers to a number of the analytical aspects of cognition and emotional-cognitive charging discussed there. The point to remember is that classifications are not neutral. Individuals and groups are stereotyped into safe, dangerous and ignorable groupings, and most often arranged hierarchically. This involves the charging of neurological knowledge units, so that in the process the spaces and boundary markers between groups and/or levels (between 'them and us', for example) become emotional-cognitively charged, sometimes quite strongly. It is these charged spaces and markers which constitute taboos, hierarchical social distancing and social boundaries and give them causative force in terms of behaviour. The degree of charge is the degree to which taboos and boundaries can be said to be strong or weak and/or social distances great or small. So, as a result, many groupings appear to consciousness as *clearly bounded*, and at least some of them seem almost *natural*.

Recognizing Social Boundaries

Some individuals, behaviours, objects, and bounded images of combinations of these, generate particular combinations of heightened sensitivity, excitation and/or tranquillity which attract us to them (desire reactions); the sight of them subconsciously pulls us towards them, there seems to be a force greater than ourselves at work; if they are clearly bounded we desire to be taken in by them. On the other hand, some objects, individuals, behaviours and bounded groups are linked to tension/anxiety (fear reactions); we come to be repelled by these, it seems as if there are dangerously charged forces (spaces) and/or barriers between ourselves and them. Thus individuals know, viscerally, that it is potentially dangerous to advance further when they are entering tabooed spaces or approaching a social boundary which they should not cross.

Individuals both variably recognize and differently value bounded groups, coming to not just *know*, but also *feel* that they share boundaries with specific others (but not different others). At the level of self, for example, parents, siblings and other loved ones often come to be felt to be inexorably linked with self within an unbreakable boundary. These feelings are very often recognized by others, making them not just individually recognized boundaries but also socially shared boundaries. In a wider context, individuals usually agree, or have previously agreed, criteria for designating, who are the members of their communities, ethnic groups, religious denominations, status categories, places of work and, today, states.

Political boundary building includes not only the development of administrative boundaries. Political leaders most usually make laws and regulations (rules) which give formal recognition and enforcement to more informal taboos and social boundaries. So, human nature motivators, interpersonal pressures, community norms, laws of states and policies of governments all give recognition and support to, for example, specific sexual practices, family forms, professional norms, hierarchical relationships and organizational structures. There is a potential problem here however. A desire for rules may serve some emotional-cognitive 'requirements' of humans (especially among rule makers) but can easily become somewhat disconnected from the directional tendencies of a number of other human nature propensities. Moreover, increasing social diversity means that no one set of rules and regulations makes everyone happy. Here there is a tendency to go to another powerful human nature propensity: the tendency to look for either the 'greatness' or danger of certain families, churches, organizations or 'countries', for example. And it is these senses of greatness and of danger, that must be as much an objective for our social science as an analysis of what taboos, rules and boundaries might exist in given circumstances.

In fact, most political (and other) leaders usually know this. Despite what they say in public, they understand that 'greatness' sells and mundane functionality is only useful as a last resort to combat very identifiable dangers. So, politicians, priests and such-like commonly use ritual activities as a means of attempting to get people to recognize and spiritually identify with the boundaries (and taboos and rules) they favour, or will make them 'great' or at least think will make (or keep) them popular (for a discussion of these approaches see, Hamilton, 1995, pp., 129-136). How might this work? During ritual activities 'positive' frantic-feedbacks are generated by the ritual activities themselves: such as dancing, listening to manic music, being held spellbound by stirring oratory, chanting, marching, singing, cheering, screaming. These independently reward individuals, intensifying identification with the specific taboos and boundaries symbolized (visualized) by the ritual activities.

Ceremonial rituals are especially common when new members are accepted within boundaries. Ideologies, too, play their part. Ideologies - and social theories - put together a limited number of variables in such a way as to make it appear that some boundaries are *natural*. When we add the human tendency to teleological thinking we soon 'discover' that some are *necessary* for the fulfilment of the human purpose. All of this, of course, also works in reverse; that is, to make some individuals, boundaries and groups seem ex-tremely dangerous. It must be remembered that some boundaries are much easier to recognize and identify with than others. Of the examples above, family, religious denomination and nation state are generally stronger than community or place of work. Ethnicity comes and goes with different his-torical events (Cf., Handlin, 1951, 1966; Brogan, 1986; Epstein, 1992; Ma-son, 1970; Hicks and Leis, 1977).

The key point is that the degree to which boundaries are emotionally felt is closely related to species typical human desires and fears. Political leaders (or the media) can be successful only up to a point in artificially creating a sense of greatness or danger if their portrayal is too far out of line with human nature propensities (or is too rationalistic). Moreover, the degree to which boundaries are accepted depends on the extent to which the dangers they are supposed to protect against exist. Community and ethnicity as boundaries, for example, are always much stronger when their members are under siege than when no external danger threatens. The same, of course, can be said about the taboos which are part of the enforcement of boundaries. Taboos, like boundaries, come and go (Steiner, 1967; Thody,

1997), with varying circumstances. Often they have to be taught with considerable vigour. Young Alexander Portnoy explains:

> "What else, I ask you, were all those prohibitive dietary rules and regulations all about to begin with, what else but to give us little Jewish children practice, practice, practice. Inhibition doesn't grow on trees, you know – takes patience, takes concentration, takes a dedicated and self-sacrificing parent and a hard working attentive little child to create in only a few years' time a really constrained and tight-ass human being. . . .Why else . . .but to remind us three times a day that life is boundaries and restrictions if it is anything . . ." (Roth, 1971, p, 88; see also, Davies, 1982).

But even with all that practice, it was possible to relax, if no one was threatenng or watching. In a Chinese restaurant the ban seems to have been lifted on pork ". . .for the obedient children of Israel. . .because. . .the elderly man who owns the place, and whom amongst ourselves we call '*Shmendrick*' isn't somebody whose opinion of us we have cause to worry about. . .[to him and his waiters] we're just some big-nosed WASP!" (pp., 100-101).

Lobster was still out of bounds, but Alexander's mother had a logical explanation: "Because it can kill you! Because I ate it once, and I nearly died!" (p., 101). Like with Mrs. Portnoy's threat of death, sexual taboos, as we have seen, are often used to enforce conformity to general boundaries (Davies, 1982; Douglas, 1970). This is because, it can be argued, they are so full of potential danger, and so easily bio-electrically charged with a variety of emotional-cognitive responses and mood states. Lust, love, jealousy and envy, for example, are often activated during sexually based relationships, often generating strong feelings that taboos and boundaries are *safety* in a most fundamental sense; if such feelings can be transferred to taboos and social boundaries *generally*, the keeping of taboos and the protecting of social boundaries, in a variety of areas, have a powerful motivational basis.

And, unlike lobster (properly cooked), a violation of sexual taboos *can* result in death – the methods ranging from stoning to disease. And a number of the rituals used to enforce them can hurt – circumcision at puberty in pre-anaesthetic times springs to mind. And 'rules' can inform fantasies – a threat of castration for adultery is not a pleasant prospect. Even in a less dramatic fashion, violating sexual taboos can lead to considerable anxiety, ambivalence and fear (not the least a fear of rejection). Embarrassment, shame and guilt, intermingled with envy, jealousy, and so on, are com-mon and can easily result in a strong sense of the extreme seriousness of the *breaking of taboos generally*. All-in-all, it seems better to stick to the taboos and boundaries on offer, not only those concerned with sexuality, but also those which regulate interpersonal behaviour generally.

Boundaries, like taboos and rules, protect against dangers. Therefore individuals who wish to get inside them will have to be motivated to alter their behaviour, or at least control their emotions, desires and fears, sufficiently to remain on reasonable terms with those already encamped within. They will most likely not only have to defer certain gratifications but also be seen to be doing so. It is here that deceit and self-deceit were probably especially significant in human evolution. Simple deceit might be used in order to try to gain advantages under the protection of group life. For example, an individual might profess loyalty, claiming maximum benefits while trying to make a minimal contribution. But that runs the deceiver the risk of being expelled, if not outlawed. When placed *outside* the law which particular boundaries provide, an individual very often can be harmed without any sanctions against the perpetrator.

To get the best from protective boundaries, in both the medium and long run, it is important to give the impression that they are vital and that they are worth defending - with one's life if necessary. The best way to accomplish this bluff is to appear to believe it – which is easy if the conscious mind is convinced that it is true. This suggests that through evolution an ability to relatively easily believe in particular group norms, values and ideas, despite having previously believed in others, developed. Charges of hypocrisy aside, this does not mean that an individual would change their basic personality - at least not very much. It would just be that the symbolic abstractions which make up ideas, norms and values would be changed in service of self's deeper desires and fears, which themselves would change very little.

This, one could argue, is a major reason for the evolution of self-deceit. That is, individual interests come to be served by a strong belief in a group's values, norms and ideas, genuinely held because of the charging of the desires and fears which underpin *wishing to join* a particular group (or to undertake particular social interactions) in the first place. Certainly, when first joining a group, given social and moral support from its members, it is often quite easy for us to rationalize our past, present and future in such a manner that the goals and professed fears of the group become part of our own holistically and teleologically created needs as well - why had we not understood all this before, we wonder?

Emotions and desires, however, are not always easy to control and the fears that lead to the creation of boundaries, or to the ensconcing of self behind some already in existence, are not so easily dispensed with. The fickleness of values, beliefs, norms and ideas can in itself become a basis for anxiety and social conflict. Therefore, humans have evolved a considerable ability to identify and prevent potential 'danger spots' within boundaries before they erupt into full blown disruption. This, as we have seen, is often through the generation of taboos. Taboos, however, usually only work effectively in relation to given bounded units. Within the *agreed definitions* of a particular group, taboos can give clear indications as to what an individual is required to do. Taboos also indicate how much, and what kind of, "tabooizing power" an individual is given - that is, the licence to declare specific property, persons and/or behaviours off-limits to all others (Steiner, 1967; Radcliffe-Brown, 1952b; Thody, 1997). From this, individuals can better know what expectations it might be reasonable to hold. In that being a member of a bounded group most often requires recognition of the rights of *all* members to certain expectations, specific individual rights - even if they are unequal - have the appearance of being derived from socially acceptable authority (if not having been given spiritual sanction) rather than being of blatant self-interest.

Having sacrificed a degree of personal sovereignty and 'freedom' when joining others within particular boundaries, humans have already threatened their own individuality and personal power. They have altered and restricted their behaviour and have probably also limited some of their potential for desire fulfilment (in non-group fulfiling ways). Therefore, often *rights,* signify both sacrifice and membership; as such they become emotionally sacred to the joining individual. Violations of an individual's rights, thus, come to be felt emotionally. They can be felt as an attack on the *value* of self, as a draining of *mana* (self-respect, sense of control). All the more reason, then, individuals feel, not only to restrict behaviour of self in order to avoid excuses on the part of others for violating self's rights, but also to support the policing of others' behaviour so that the possibilities for violations do not arise.

At the same time, in so far as individual rights are only definable in terms of the relationships of individuals to each other - as agreed by all of them (a bounded group) - an attack on an individual's rights is an attack on all of the members, and as such on the boundary itself (agreed protection of rights for all). Such attacks often generate an overwhelming desire for vengeance among members of a threatened group. Attacks on groups, or threats of such, also generate a strong sense of exclusiveness, or at least a desire for such, among members of groups.

Processes of Inclusion And Exclusion

In Chapter Five we considered a number of the powerful emotional-cognitive processes which are involved in the creating and charging of the species typical knowledge units related to the taboos, rules and social boundaries discussed above. It is worth reminding ourselves of these. They are: lust, love, jealousy, hate, envy, the control of envy, empathy, guilt, embarrassment, shame and depression. Each of these processes entails the development of charged - both attracting and repelling - spaces and/or boundaries between self and social others. Images linking self with certain others, objects and behaviours, come to be connected with heightened sensitivity and excitement mood states. As a result, an attraction between parts of the image - that is, between self and others and between self and specific objects and behaviours - is generated. Other images, however, connect conglomerations of specific individuals, objects and behaviours to anxiety/ tension mood states; these images act as repellents between self and those individuals, objects and behaviours. All of the emotional-cognitive processes listed above, when cognitively elaborated, thus motivate the social processes of inclusion and exclusion.

For example, lust separates the world into those one physically desires and those one does not. Love clearly separates the worlds of people and things into two clear and very distinct groups: those one loves and all others. The first group is very small and receives considerable devotion, loyalty and effort from self. The second is very large and usually only receives the above in terms of pre-determined reciprocal arrangements. Jealousy divides people into those who 'belong' to self, those who would like to steal these away from self and neutrals. Hate designates a clear category of people whom self would like to hurt, attack, avoid and, above all, exclude. Envy designates those objects, behaviours and relationships which self would like but about which self must be very cagey indeed if self is not to arouse the jealousy and hatred of those individuals who now possess them. The human capacity for empathy sets apart those characteristics and individuals one is willing to trust, to be loyal to and to feel safe with; that is, those with whom one is willing to form an alliance. In general, those who develop similar notions about co-operation, expectations and the dangers involved in certain activities will develop at least a degree of empathy for each other. These, usually, will find inclusion with each other relatively easy to achieve. Those who seem to have different notions about co-operation and different expectations, and who even seem to be potential dangers, will be excluded.

But in social interactions we usually want somewhat better guarantees than those merely provided by empathy. As we have seen, taboos, rules and boundaries provide a more or less formalized method of establishing rights so that every encounter between individuals is less dangerous than it might otherwise be. The establishment of rights designates, ahead of time, the amounts of jealousy, envy, vengeance and general aggression, for example, an individual can expect to get away with. But generally, as noted, these rights only operate effectively within the confines of a bounded group. The development of rights thus tends to reinforce notions of who is included and who excluded. These processes of inclusion and exclusion can become more or less formalized ranging from the informal criteria of inclusion found in children's gangs to the relatively very specific criteria of professions or militaries or citizenship, for example.

These examples introduce another means of maintaining group solidarity; that is, through the development of a strong sense of duty. The development of relatively elaborate desires for inclusion, for whatever reasons, is most often reinforced by guilt, embarrassment, shame and depression for not living up to group expectations. This is quite apart from a fear of punishment which might follow transgressions. These more 'negative' emotional-cognitive processes punish transgressions, often unmercifully, by emerging in dreams, daydreams and fantasies. They become stored as memories and are activated by the slightest flicker of stimuli which activate knowledge units and mood states linked to a sense of dereliction of responsi-bilities. They provide a firm basis for excluding those who do not live up to group norms.

Such 'negative' emotions, on their own, however, would be unlikely to work to maintain social boundaries for extended periods of time if they were not part of a larger package. This larger package includes spiritual highs derived from being included, loving and being loved, gaining desired objects or persons, feeling deep empathy for others and being members of exclusive groupings. In other words, strong feelings of camaraderie and fellowship develop which not only provide a basis for group solidarity but also for the inclusion of those who fit in and the exclusion of those who do not. Humans clearly have a strong propensity for social bonding and boundary building.

The development of *definite criteria* for inclusion (in both loosely inclusive networks and tightly bounded groups), however, generates certain dangers of its own. For example, due to the human tendency to continually reduce variables to a few stereotypical images, humans only understand in complex detail those things immediately important or intimate to themselves. This means that feelings of being *deprived* are very often, in fact, generated by comparisons with *included* individuals (on the concept of 'relative deprivation' see, Runciman, 1966; Davies, 1969; Gurr, 1970). It is usually those who are expected to live by the same taboos, rules, standards and notion of rights that self has accepted - but who appear to cheat - that generate senses of deprivation and injustice among humans. Because individuals come to expect rights which are measured relative to those with whom they have shown considerable empathy and camaraderie, especially in terms of shared notions concerning deferred gratification, it seems a major violation when such others do not control their own desires or envious feelings.

It does not seem a major violation of individual rights, for example, when a movie star (a distant, almost mythical, stranger) gets paid three million dollars for a few days work, but a complete injustice when a mate at work gets promoted ahead of self if self feels they do not deserve it as much as self does. For such 'victims' - who have probably greatly controlled their own envy towards their mates over the years - empathy and feelings of camaraderie quickly disappear and envy expresses itself in terms of a great sense of injustice and indeed, moral, and sometimes physical, aggression - usually accompanied by a strong desire for vengeance. By the one promoted, moral indignation may well be expressed at the envious and aggressive behaviour of the 'victim'.

Humans have a very long standing evolutionary basis for developing a number of fears concerning the potential dangerousness of close others. During the period when humans were evolving, close others were the potential competitors, allies, sexual partners, friends, enemies, providers of resources and/or protectors. It was these who might violate trust and/or let down, cheat, humiliate and degrade self. So humans have evolved a human nature which keeps an 'emotional eye' out for such potentialities from close others. And, moreover, they want to control and punish individuals who cause such dangers. Certainly, throughout history humans have established taboos, rules, subordinations, expected deferences and symbols of respect and honour which they expect others, especially significant others, to comply with. These have been enforced largely through gossip, parenting, public opinion, ritual preaching, worshipping and personal invective.

HIERARCHY, POLITICS AND SPIRITUALITY

It is for this reason that humans often seek to establish predictable, *guaranteed* hierarchical relationships *with regard to those nearest and/or dearest to them.* And so humans usually politick and police most vigorously and personally hierarchical distancing, taboos, rules and boundaries which involve themselves and their significant others. Humans want their children to treat them with a certain degree of respect, especially in public, and to reflect well on themselves in society; humans want colleagues and bosses to treat them with dignity. It is relatively easy to accept the privileges of a powerful patron, or the orders of a boss or of a field-marshal, or the preaching of a preacher so long as they are given according to known rules; and that all taboos concerning the avoidance of humiliating references and/or behaviours and situations which might threaten the dignity of either party are observed. Joking relationships can help greatly here; as can a sense that the patron/boss/field-marshal/priest represents a higher, almost 'spiritual' order. In other words, humans usually do not mind having less than someone else so long as it does not reflect badly on themselves; so long as it is based on the rules of the game, so long as it is not humiliation for self. They may even think that it is fair – that their job is harder than mine, or it carries more responsibility, or that the holder of power or eternal Truth has been granted special spiritual qualities, and so on.

Humans, then, evolved both fears and desires for enforcing and adhering to procedures which control the behaviour of close others. As humans began to live in more dense population concentrations, and to be involved in more specialized divisions of labour, fears of being violated by close others were no less for it. It is here that hierarchical evaluations and stereotyping come into play. As we have seen in Chapter Six, when humans evaluate their environments they do so in a hierarchical fashion. All individuals, presentations of self, behaviours and objects are not equally valued. Behaviours of violating trust, cheating, humiliating and degrading others result in a socially accepted devaluation of those behaviours and those individuals who practise them.

As a result, human nature motivated (and personally enforced) sanctions become socially recognized (and enforced) as a *hierarchical* ordering of behaviours and individuals. When this propensity operates in conjunction with the human nature tendency to stereotype, they result in a sort of hierarchical arrangement of relatively small, more or less bounded, 'social worlds'. Generally, humans police their own, but 'live and let live' with regard to a multitude of other such stereotyped social worlds. Linkages between them are based on often rather tenuous (from a social determinist's point of view) legal, political, voluntary and/or happenstance connections. But, most likely, this is how 'political society' works.

Hierarchy

More than many other social processes, however, social hierarchy has been considered to be in some way a *structural* part of a larger whole - society - separate from any notion of laws or governmental authority. This is, perhaps, because it appears to encompass everyone. It seems impossible to personally escape from its effects. With the development of enlightenment sciences this remained the view. For Durkheim, and even more for functionalists such as Spencer and Parsons, for example, it was tied up with a functional – implicitly rational, hierarchical – division of labour. Although this approach left a degree of ambiguity (was the division of labour the driving force - Durkheim, Spencer - or the hierarchical relationships – Parson's system's imperatives? Which causes which?), the notion that it was a rational, functional part of an evolving social whole, which was making progress, was not challenged. A complex division of labour was deemed to be essential for advancing industrial societies, so it was assumed that not only was hierarchy inevitable but that, in the end, it was for the benefit of all.

From at least after the enlightenment, personal inequality itself was never advocated, other than as a short term necessity. It was not to be like the days when it was believed that every person was in a God ordained place and that they should not challenge the order of things. Individuals were to be free to improve themselves, to work hard, or to study; to do whatever was necessary to advance themselves. The 'system' was not unfair; some people might behave badly, but that did not make the system wrong; above all, it did not make it any less a system. Indeed, there were those in the social sciences who set out to make hierarchy into even more of a specific *structural* aspect of 'a larger whole that was society'. Marx, for example, set out to show that, as a system, capitalism was made up of two relatively distinct classes, owners and workers. The system/society gave each of these very different interests, which, as the system matured, individuals would increasingly become aware of. The workers (proletariat) would recognize that they had been exploited, *by the system*/society, and being by far the most numerous (quite apart from being in tune with inevitable history) would overthrow the system and introduce a new one, socialism.

Here hierarchy is presented as two antagonistic classes, but classes nevertheless which make up a system of society. Others, less critical of evolving industrial society, also used the concept of class. Some of the early American studies (Cf, Warner, 1960: Warner and Lunt, 1941, 1942; Lynd and Lynd, 1937; Dollard, 1937; Sorokin, 1947) got out of hand in terms of an analogy with geology where 'natural occurring' strata were looked for in the assumption that they would tell us something about the nature of evolving society. These ended up with multitudes of strata (classes) which became not much more than descriptive categories. Some sought to use occupational categories (Lipset and Bendix, 1959; Inkeles and Rossi, 1956; Blau and Duncan, 1967; Brown, 1977; Goldthorpe, *et. al*., 1980; for a discussion see Hamilton and Hirszowicz, 1993; Marwick, 1981),which told us a lot about patterns of mobility and occupation change in industrial societies, but not much about the *processes* of hierarchy. Increasingly, scholars were unable to identify classes on the basis of any clear criteria (Hamilton and Hirszowicz, 1993) as differentiation between manual, white collar and professional increasingly blurred.

The problem, it can be suggested, is that all of the above were approaching the issue from the wrong direction. There was the assumption that somehow inequalities had to be part of 'the *system*' of particular societies rather than the result of normal human nature processes. This last suggests that some of the very processes which make up what we call society emanate from human nature propensities to hierarchy (as suggested at the end of the previous section). If we consider the picture of human nature developed in this work, this approach makes sense. As we have seen, humans, guided by human nature, recognize and specify dangers and opportunities in social life. In the process they inevitably evaluate their physical and social environments.

During the development of self-identities, *charged* hierarchical evaluations of self in relationship to others are established. This is more or less confirmed by social others through the types and amount of attention, rewards, and prestige directed to self. Shared hierarchical evaluations become more official social processes when they result in formal procedures for differential allocations of attention, material rewards, social honour and authority. These include numerous generally agreed social distances, taboos, rules, duties, obligations, demands and boundaries. This provides a degree of predictability in, and some control over, an individual's life circumstances. Formalization takes place through politics. Human nature-influenced evaluating, thus, becomes social processes of hierarchy through politics.

In terms of the perspective being developed here, we must remember that social hierarchies are not 'givens' which remain stable over long periods of time. Rather, hierarchies grow organically; they are always in the process of becoming, parts dropping off, parts being added, changing shape, criteria of evaluation always evolving, and so on. When, for example, the number of individuals increase in an area (increasing the 'dynamic', 'moral density' - Durkheim, 1964, 1982), the more difficult it is for many individuals to fulfil traditional expectations. Individuals can be excluded from access to social attention of the kind they had come to expect, or have been accustomed to; they feel deprived, possibly violated; they seek a new role - redress; they seek to preserve their status and to maintain reproductive opportunities; they seek change; others react; change is inevitable whatever the specific outcome.

Recent technological developments, and greater and greater population densities, have, in fact, made it possible for humans to generate, sell and buy a multitude of new – ever changing - toys and behavioural games (professions, life styles) to participate in as part of their coping with the human nature propensity to hierarchy. At the same time, the human nature tendency to escalate expectations (that is the propensity for the elaboration of desires for 'improvements' in life circumstances) feeds the human fear of failure, and fear of being humiliated, generating a need for *success* (the Bitch Goddess, herself, Cawelti, 1965); and so the games of hierarchy go on. There can become a compulsion to possess more and more toys to display, more and more professions to belong to, more and more activities to professionalize, more and more cultural snobbiness to ritualize and flout, more expensive wines to offer, more lofts to be turned into flats, more 4-wheelers to cow Ford Fiestas, more potential lovers to seduce, all to stay even in the games of society; and all due to propensities which gave reproductive success in times past.

In the processes of hierarchy, a social order of sorts is established. A social order which might not be the ideal from an enlightenment thinker's or an engineer's point of view. But a social order nevertheless. It works, like my liver. How do I know if it is the best one I could have? And does it matter so long as it works? It is a social order which gives humans something meaningful to do; an order which provides purpose in the lives of multitudes. And all of this drives social evolution. And it drives politics. In fact, hierarchy has increasingly become a *political* issue. As individuals struggle for inclusion and others resist, the rhetoric of hierarchy has increaseingly become the rhetoric of politics. There is a tremendous fear of losing out. And given the absence of traditional - and the precariousness of many current - demarcations, it not always easy to know where one is in the race of life. Desire fulfilment is often a never ending process, and feelings of envy seem never to go away; there are constant fears of personal unworthiness. Guilt easily rears its head because there is always a sense of not having controlled self, and conversely, of not having contested sufficiently. It is little wonder that the cry for equality has become a major political demand in all modern societies. But the cry really means 'make me not a loser'; 'give me the guarantee of *recognized* success'.

And so from all the reactions to these fears and attempts to fulfil desires, there arises a more or less functioning division of labour. And in the process, individuals, institutions, categories of behaviour and ideas are ranked in such a way that most individuals have some notion of who and what is safe, dangerous or neutral; of who might provide what kind of protection and who might be seeking protection; of who has what to sell; who might want to buy what self has, and so on. In other words, hierarchical relationships provide a basket full of rules, guidelines, taboos and protective shields in the dangerous worlds in which self-aware creatures tread. In this sense, hierarchical propensities (and resultant natural kinds) provide a major ingredient of all human social orders (political societies).

Nevertheless, hierarchy on its own can remain dangerously ambiguous. But it is supported by a number of other behaviours (and their motivators), including the binding/bonding and boundary building behaviours discussed above and in previous chapters. It can be argued, however, that its major support comes from the *political* activities of judging, administrating, adjusting, creating, defending and discriminating among the rules, restrictions and social boundaries which make up the varieties of sexual patterns and other behaviour patterns which can be argued to be species typical of humans. Politics also play a major role in the processes of inclusion and exclusion, discovering good and evil and in generating social evolution; they have their own contribution to make to the human story.

Politics

Politics is about competing, conflicting, following, joining, subordinating, dominating, negotiating, influencing, deceiving, cajoling, condoning, condemning, helping, trusting, advocating, bargaining, convincing, balancing, trading off, agreeing, and so on, These involve *power;* that is politics represent relationships in which attempts for influence, control, clever domination and creative subordination are commonplace. Politics is where the 'human will to power' (desire to enhance and protect self) expresses itself in the context of power networks. Taken out of context, politics seem to defy many of the enlightenment assumptions about human rationality and the pur-suit of scientific administration. It seems sloppy; to be about lying, greed, large human egos; politics does not represent behaviour we wish to encour-age. Therefore the art of politics has not figured large in theories of human social behaviour. Yet, politics (not administration or management) can be argued to be among the most significant of processes concerning the work-ing of human nature in the larger social sphere; this is especially the case as human societies become more dense of people, complex of technology and their members less dependent on purely physical survival.

Politics is clearly much more than just activities concerned with inclusion and exclusion, or just about administering rules, taboos, boundaries, and hierarchical relationships. But we must not underestimate these. A major 'job' of leaders or governments (a natural kind) is to judge among competing claims, enforce judgments and to generally protect people. If politics do not result in these, governments do not long endure. Nevertheless, all of this is a product of reacting to the dangers inherent in social life, not a product of some sort of human rationality or of some grand design of history. Considerable social stability, predictability and agreed social relations are established through politics. Complex divisions of labour are established and maintained through governmental decrees, policies and laws; but these are a by-product of human nature, not a cause of it.

Politics, then, are, at base, about the games humans are motivated to play in attempting to present self in a favourable light, to form alliances to share gossip and to steal (and protect) goods, sexual access and political control (while accumulating status); they are about war; they are about *winning* elections; they are about balancing behaviour in the face of the human dilemma between being engulfed or expelled. Almost the whole range of emotional-cognitive processes, desires, fears and advanced cognitive processes are involved in politics. Vengeance seeking leading to reasoning about justice; obsessions, addictions, self-deceit, they are all there; love and powerful identifications with heroes, stereotypes of enemies, worshipping of symbols and leaders; envy, jealousy, depression – none of them missing. Active politicians put self on the line; self has to come out winner; self faces humiliation, defeat, depression. But stereotyping, rationalization, strong identifications and teleological arguments all help protect self.

So, we hate politicians because they represent the worst of us; they highlight our weaknesses, they show up our lack of rationality. But we look to them for solutions to our dilemmas and problems. We try to separate statesmen from 'politicians' and tyrannical leaders. Statesmen are kind and honest, politicians and tyrants are dishonest and only self-serving. In our imaginations and stereotyping, politicians and tyrants get worse and worse, statesmen get better and better. Before long, statesmen become symbols of the polis, they become cultural heroes, if not gods. As a result, image becomes more important than reality – no, the image becomes reality. And the statesperson has to live up to the image. So image management, and creation, becomes as important as day by day administration.

This is not new in human history. Pyramids, temples, splendid palaces, lavish court life, 'the White House', political conventions, the English Monarchy, 'the Kremlin', did not come into existence for nothing. These may all have provided a great deal of employment, but so too could have digging holes and filling them in, or making a mountain in the middle of the Sahara desert. (The Disney Corporation may think of this before too long.) They are a reaching to higher things, to greatness, to the heavens. So, politics easily comes to be a locus for defining good and evil, for setting out human goals; for giving us a higher calling. Legendary leaders of great courage and wisdom are the stuff of mythology. Politics, then, ranges from the work of the Devil in his intrigues to secure the triumph of evil, to the work of God, seeking the victory of good. And with this dimension of politics we can see another powerful propensity of human nature, that is the tendency toward spirituality.

Spirituality

Human spirituality is an emotional-cognitive 'high' in which charged holistic images of a higher human purpose are generated. Brain opiates undoubtedly contribute when individuals focus on the symbols and people visualized in such images. Human spirituality creates an image of harmony where tranquillity, direction and purpose have their own reasons for existing. This has made a considerable degree of sociability possible for humans. If people delude themselves that they are altruistic, the more likely they are to act in socially helpful and altruistic ways (Badcock, 1986); the more individuals believe in a higher purpose, the more likely they will live for it.

We can use an extremely general definition of human spirituality. Spirituality can be taken as a feeling that there is order and form (design) in the universe, and that this means purpose. Religions, of course, fit this definition, but so does much of philosophy, science, political doctrines and everyday thinking. Spirituality implies that through the right prayers, rituals or knowledge, self can be purified, if not perfected. Human spirituality seeks to discover universal design, purpose and meaning as a path to social unity, in order to further create the conditions for human perfectibility, peace and general happiness. Spirituality gives humans very intense bio-electric highs which provide for deep concentration which often becomes totally absorbing.

Spirituality often includes a recognized subordination to a universal design or purpose. This can give individuals a sense of purity and security, the security of believing that all will be well even if this is not currently the best of possible worlds. This can imply causation and a sense of power, because if design and purpose suggest direction, and if humans are in tune with this (are in the right state of grace/purity, have the right knowledge, are dedicated to a higher purpose) they have an explanation of the why (cause) of their own behaviour: to fulfil God's, History's or the Grand Plan's destiny; and at the same time, they have a sense of power: they are the *cause* of the Grand Design coming to fruition. So, in a circle, there is purpose and direction in the universe and humans are the objects, causes and evidence of it. To be sure, the supernatural nature of spiritual power can be negative. Witchcraft as a supernatural power for harm is an example (Cf., Kluckhohn, 1944; Evans-Pritchard, 1937; *Malleus*, 1486; Anglo, 1977; Douglas, 1970a; Lea, 1957). In modern societies mental illness is granted the power to cause individuals to be dangerous (Nunnally, 1973; Grove, 1970). Foreign ideologies with 'brainwashing' potential are sometimes also credited with just such power.

The search for the nature of the Grand Design and the direction of the human purpose motivates humans to adjust environments and presentations of self in the face of reproductive and survival dangers. Dreaming and daydreaming about purpose, and seeking to understanding it, generate expectations, rights and ideologies, through the bio-electrically rewarded process of mentally making self safe via powers which are treated as having the ability to control the universe. A number of subordinations are sought out, subordinations which can be flexible in terms of dealing with real individuals, but quite binding in a contractual sense. (An individual can leave a church because the preacher is 'no good', but keep a contract with God by still believing and looking for another church.) Theoretical notions of universal trust, loyalty, mutual obligation, obedience to duty and altruism, for example, become highly desired 'qualities' through the human propensity to spirituality. These are qualities to which we often attribute an almost supernatural existence separate from humanness. This process removes ambiguity from a number of social relationships, providing a degree of safety through standardized patterns for 'debts' and obligations. It underpins a number of natural kinds, quite apart of that of 'church'.

IN SEARCH OF NATURAL KINDS OF BEHAVIOUR

Nevertheless, it is clear that there are no perfect, or completely harmonious, procedures for creating taboos and rules, for social bonding and boundary building or for establishing hierarchical distancing and the regulation of behaviour - despite the human propensity to spirituality. There are most often too many conflicting emotional-cognitive motivators, desires and fears operating at the same time, some of them decidedly non-spiritual in orientation – and anyway, one person's spirituality is another's danger. So, politics is an inevitable part of human social life (and what is more, we have a propensity for politicking). Politics/politicking (as being largely about: ministering to rules, taboos, social distances and boundaries) iron out and cool many of the conflicts and ambiguities in a number of other processes and as such go a long way towards keeping general desires, lust, hate, envy and jealousy under control. Or, at the very least, it can be argued that politics tend to control the behaviour which might emerge from excessive desires and too much expression of such things as envy and jealousy.

There are many more propensities which will have to be discovered in the development of our social science. For example, in Chapter One, and variously throughout this work, it has been suggested that humans have a strong disposition to feel that there is purpose in their lives (and in the universe generally), for generating abstract ideas/beliefs, for co-operation, for being social, for 'altruism' and for accumulating knowledge. There are undoubtedly more which have had little or no mention here – such as a propensity, for example, to rear offspring - which we will want to explore in our search for natural kinds of behaviour.

Also, besides the natural kinds discussed in this and the previous chapter, our social science will have to subject a number of other long discussed human social behaviours to the same analysis of 1) seeing if they have sufficient human nature underpinning to argue that they are natural kinds of behaviour and 2) seeing if there is enough anthropological and historical evidence of their universality to support 'positive' conclusions from the first task. Human sexuality, for example, must be fully considered in this light, as must a number of other 'social boundaries' – from 'self' to 'the state'. Of special interest must be an examination of: *specific* male and female sexual identities, self-identities, monogamous relationships, nuclear families, extended families, kinship patterns, bands, 'communities', patron/ client relationships, classes, castes, voluntary organizations, formal organizations and states.

It is highly likely that, despite the fact that a number of the above have had pride of place in one or another of previous teleological explanations of human behaviour and/or desires for human perfectibility, an analysis (in the manner suggested above) will present a very different picture. I have already argued, for example, that the notions of 'social structure' and 'social systems' are teleological and not of much use in an attempt at a realist theory of human behaviour. How many of the above are dependent on such notions rather than having a real basis in evolutionary theory and a consideration of human nature? For example, as I argued in Chapter Two, it is likely that the powerful notions of extended families and kinship may have more to do with human optimism and sociobiological theory than is warranted from actual observations or a consideration of human nature.

On the other hand, how many behaviours have a solid basis in human nature and anthropological and historical universality, but have been discarded by social scientists because they do not fit in with rationalistic and teleological assumptions? Instead, they have been treated as simply deviant (thus not fundamental to human social life) or as non-rational behaviour (thus not really 'human' behaviour); in either case as not being worthy of serious consideration when explaining something as 'complex' and perfectible as human societies are meant to be. We have seen, for example, the importance of human vengeance seeking, yet most theories of society or politics ignore such behaviour more or less completely or, at best, treat it as deviant, non-rational behaviour. Similar behaviour when seen in modern societies, such as found in playground cliques, juvenile gangs, criminal gangs, terrorist organizations, vigilantism and conspiratorial factions in organizations and in politics, is equally ignored in theories of society as not being worthy of being part of an explanation of society, *qua* society. Yet, almost all major organizations, political parties, military structures, revolutionary movements and (in may cases) modern states, are run by, if not organized on the basis of, very small cliques (however temporary) of insiders who are bonded together on the basis of mutual trust, friendship, shared senses of danger, shared enemies and a tight inter-dependence (see note five, p., 234) This is quite apart from the impact of, say, exclusive religious cults throughout history (including the Jesuits and Masons, for example) on the organization and evolution of social behaviour (Cf., Evennett, 1968; Cohn, 1970; Willis, 1970; Wilson, 1990).

Other examples of generally ignored possible natural kinds include behaviours described as: bands, voluntary organizations and patron/client relations. These are most often seen as humans behaving in a sort of pre-rational age, in an age before humans realized that they could become *organized*, or in which altruism is limited to the voluntary efforts of a few (exemplified by church fetes), or in which a greedy and evil minority take physical advantage of the superstitious, downtrodden many. Yet political parties (not even mentioned in the US Constitution) are voluntary organizations in which patronage is basic and band-like factions abound. The importance of political parties in the 'structure', running and character of modern societies is, of course, enormous.

With regard to patron/client relations themselves, godfather politics are widespread throughout the developing world and seem to spring up very quickly in more developed areas where state governmental power has either disintegrated (for example, at the collapse of the old USSR), or does not, or cannot, reach (for example, during most of the Twentieth Century in Northern Ireland, in the criminal worlds of the USA and Europe, in international drug cartels, in local governments and, as noted, within political parties). This is quite apart from their long historical pedigrees in those primitive societies just past simple hunting and gathering, and in the various patterns of feudalism throughout history. What about: 'social worlds', micro-hierarchies, sports fraternities, rendezvous-like caucuses and conventions? A 'war community' (if it is different from gang fighting or patron/client based wars), perhaps, and friendship networks are further possibilities of ignored natural kinds.

It is noticeable that the problem of identifying (and getting acceptance of) propensities and natural kinds useful for social theory becomes especially difficult when we set out to explain behaviours beyond the range of close personal contact. In parenting, sexual relations, mutual love or close personal friendships, for example, we think we can understand human motivations-propensities. Once we go beyond these personal contacts, however, there is a tendency to jump to rationalistic and/or structural concepts. But, in looking for the natural kinds which make up larger political society, we cannot abandon a search for the human nature propensities (conglomerations of desires, fears and basic emotions) which might underpin them if we are to avoid teleology.

Hopefully, our consideration of taboos, rules, boundary building, processes of hierarchy, politicking and spirituality (as being human nature generated tendencies) has taken us some way beyond face-to-face contact without using the grand circularity of jumping to macro concepts, and then coming back from them to explain behaviour as fulfiling the *needs* of that macro structure. Yet, the notion of the overriding importance and greatness of the *whole* dies hard in both popular consciousness and in social theory; in both it is in opposition to the selfishness of individuals or the dangerous sectarian-interests of some parts of a whole. It is partially from this propensity, it can be argued, that in many circumstances the *polis* – or political community - (or in modern times, the state) became a major focus of individual directed attention and loyalty, and as such became extremely important as an organizing force in humans' social life.

And this is something which clearly pre-dates the modern state (Cf., Plato, 1974; Aristotle, 1885; Hobbes, 1968: Machiavelli, 1950). The polis, it seems, may well be a natural kind which easily emerges in many circumstances. To argue that the polis is a natural kind, however, will most likely require some sort of argument that it is additionally motivated by a number of the other tabooizing, distancing, avoiding, rule making, personal binding, boundary building, hierarchy generating and politicking emotions, desires and fears discussed earlier. Such an argument can be made. But this is not sufficient to explain why humans can become so *emotional* about a state, or even its symbols, and why they will die for it, and kill for it, and torture for it, and why humans so often look to state action to solve all problems, and to lead us to conditions in which individual perfectibility is, if not automatic, pursuable, and progress is more or less assured. Clearly, the human nature propensity to spirituality (in which the whole of existence – God's, the Universe's or the Future's - is perceived as being greater, more lofty, than individually felt desires and fears or the mundane existence of the here and now) is a major force in human social behaviour.

In this and the previous chapter, then, I have suggested a number of human nature propensities/inclinations, and some possible resultant natural kinds of human social behaviour. It has been noted that clearly there are additional ones which will need to be elaborated through increased understanding of the neural sciences, human evolution, psychology and history. Some of the natural kinds considered in this chapter may benefit from being fused, others split into more than one (perhaps states, for example – between centralized and decentralized); some natural kinds will, when more evidence is gathered, have much greater support than others, some may have to be dropped. It will, of course, be necessary to analyze each natural kind - once it is firmly established - in terms of how it works and how it does, or does not, link with other natural kinds. We will want to know how various combinations result in a variety of versions of political society. We will want to know some of the consequences of each, and so on. Do, for example, certain combinations of human nature motivators push us towards a de-centralized state and others to a centralized one?

But this is the way of science; materials and events and processes are observed, categories created, theorizing engaged in and conclusions reached. These are not once and for all conclusions, they are tentative conclusions. But they are conclusions which are based on a belief in science, a belief that human nature exists; a human nature which is not teleologically derived, but rather is seen as a number of processes, evolutionarily derived, which provide a degree of consistent humanness. It is on this that we will have to base our conclusions concerning human social behaviour. This will be the next major step in developing a science of human behaviour; but it is a major step and runs beyond the scope of this work. Moreover, we still have not established fully all of the most likely natural propensities/tendencies, let alone natural kinds which will make up the social sciences. Nor have we elaborated their specifics. Still, one hopes, the door to such work is increasingly being opened.

CONCLUSIONS: A SCIENCE OF BEHAVIOUR

Human nature is a number of processes in which pre-knowledge and the charging capacities of the nervous and endocrine systems impact upon species typical analytical/cognitive processes of the brain. A major result is that the universe of peoples and objects is reduced to a few stereotyped participants in episodes of social action. These people and objects are often highly charged with significance during the process. Not the least of these stereotypes is the one of self, in which a multitude of biological processes, emotional-cognitive feelings, conflicting fears and desires, and an assemblage of experiences, are reduced to an identifiable, relatively simple, object ('me').

Self is the locus in which specific human nature motivators become a social actor. As such, it is an interaction of a particular bundle of genetic messages and experiences. When dreams and daydreams include an image of 'me' as separate from others, and from the external world, we have the evolution of self-awareness. Self, as an example of a natural kind, is based, in significant part, on a human propensity to create a social boundary around 'me'; self is the first boundary as it were, the one which separates the perception of 'me' from a perception of others. The boundary around self is precarious, changeable, but at the same time self is tenacious in its conservatism. Self is deceived and deceives; self is often passive but always a potential trouble maker; self is both the product and maker of history. Self is not a thing, fixed in time, it is a process; a process which takes human phylogenetic history into the social/political arena. As such it is the point upon which both Darwinian selection and social selection (through history) operate.

With the evolution of self-awareness humans developed both a cognitive capacity and inclination to classify others and objects in relationship to self. Self is thus classified, and self-analyzed, in terms of self's relationship to the social world. Through a propensity for holistic, hierarchical evaluation, self becomes *self-identity* and *self-esteem,* coming to include a particular mixture of latent emotional-cognitive reactions, desires and fears (making up a conglomeration of capacities, tendencies and propensities). When objects, activities and other people are classified, they are also, inevitably to a greater or lesser extent, analyzed, fantasized and stereotyped in terms of *abstract measures*. In the process, neurological inter-connections which represent symbols, concepts and beliefs about self in relationship to others are included in self-identity. Sometimes they are organized into larger, more comprehensive fantasy and/or ideological patterns. Thus the abstracted results of daydreaming and fantasies become part of the biological make-up, both unconscious and conscious, of an individual.

Self, however, does not exist on its own. Holistic images of self always include links with significant others. Emotional-cognitive processes, desires and fears which motivate care and attention seeking, thus, make humans dependent on incorporated social others for confirmation of self. Humans, then, are social animals partially at least, because of incorporating others into their own self-image requirements and depending on them for confirmation and acceptance. These others partially confirm (love, bestow prestige), or do not confirm (hate, reject), such identities (Laing, 1965; Goffman, 1971; Rosenberg, 1979). Parents are especially important in this respect because it is here that individuals first come to know who they are relative to others and also to set a value, relative to others, on themselves.

As discussed, one of the major characteristics which evolved in response to the envy inherent in human self-awareness is our emotional-cognitive capacity and propensity for developing empathy. As we have seen in Chapter Five, empathy is an 'emotionally' charged cognitive process which makes it possible to guess more accurately if others are likely to be of value to self. It makes it possible to avoid a number of dangers to self and to alter presentations of self before they alienate too many others. Individuals identify with those with whom they develop feelings of empathy. It is these who become, as it were, part of self. Feelings of empathy thus mean that a considerable amount of altruistic behaviour will be extended to those with whom one empathizes.

The behaviour of self is, thus, prescribed by species-typical capacities (pre-knowledge influenced human nature propensities) for developing self-images and feelings of self-worth. Self has the political job of, to some extent at least, controlling, coordinating, excusing and justifying species typical emotional-cognitive motivators. But self is far from competent in this and, because of the other more emotional dimensions of human nature, a long way from autonomous. As a result, while self is the actor in species typical patterns of social behaviour, only a limited number of taboos, rules, boundaries, hierarchical arrangements and ideologies are viable given the constraints imposed by human nature. Yet, even these are vulnerable to chance and to often unpredictable conflicts within human nature.

So, for a science of society we cannot depend upon an enlightenment inspired vision of human rationality striving and driving for perfection; perfection defined as self-actualization, a perfectly balanced self, a wonderfully, and completely, functional family, an engaging civil society, a synchronic 'social system', or history progressively moving to utopia. But we do not need to. Human reproductive behaviour, with all of its personal and social trappings, has existed without any of these being achieved, and will continue to do so; history is history, not the unfolding of some grand design with perfection as the end point. And it is history we wish to understand. And history is about reproductive conflicts, sad and bad feelings, emotional highs and times of despair; vengeance seeking, politics.

Yet the search for perfection does have its place in a science of human behaviour. The above utopian characteristics, in fact, are the product of a human imagination, fuelled by a powerful capacity/propensity for spiritual feelings. It is these which in so many ways are the driving forces behind the human *quest* for civilization. If decorating self is fun, and it attracts attention from objects of lust, why not do more of it? Why not do more than 'improve' the external body, why not go inward and 'improve' the core, the inner self, the soul, one's personality, make it good and pure – that should bring in very trustworthy mates. If control of self, of significant others, of circumstances, reeks of strength of character, of the power to protect, and is emotionally rewarding, why not master techniques of self-control, sacrifice, humility, confidence, assertiveness and leadership? And if talking brings attention, and love, and help, why not get good at it, why not become a singer, or a storyteller, or a preacher or a salesman? If showing off brings attention and a good evaluation and friends and sex and children and worship, why not go all out to be good at sports and/or to accumulate objects and achievements for display?

If self seems rotten, inferior, impure, sinful - naughty images dancing freely in consciousness; unstoppable, lustful desires streaming stronger than rational power - why not hide all this from public view? Keep them to fantasies. If subordination seems to verge on humiliation why not justify it in terms of commitment to a higher purpose: to God, or the cause, or the Party, or the movement, or 'one's country'? If showing off seems a bit pushy, why not justify it in terms of 'hard work', 'achievement' and 'success'? And if this is all too confusing and anxiety generating, why not support taboos, rules, boundaries, hierarchies and political practices which seem to bring some order - safety - out of confusion; which might mean that self does not have to constantly run around like a deranged maniac but can even rest some of the time?

Why not indeed. There is no reason why not. Utopia or perfection may not be at the end of the road, but who really looks to the very end of the road (which, in any case, is death, according to J. M. Keyens); who really cares if there is no utopia if one has a sense of at least some achievement derived from various combinations of friends, a secure home life, a bit of decent sex (scope to fantasize about some even better to come), a certain amount of material and/or job success, some nice daydreams, offspring carrying self into the future and no immediate sense of doom or gloom? Our science of human behaviour must take these questions, and the answer, 'indeed why not', as its starting point.

But this raises a whole new set of issues. Is that all there is to it? Might not this sell humans short? Might this not be a slippery slope to more dangerous thinking? While Darwin may have undermined the search for purpose, ultimate good and grand design in our analysis of nature, Freud made the bold assertion that nature would hide this from us, or sublimate it into religion and 'civilization' and all sorts. Nature may be a bit aimless, purposeless, perhaps even red in tooth and claw, but if it has had the decency to hide this from us, to drive us to civilization, why dig it out? Why, many critics have wanted to know, do we want to find out how the seemingly 'bad' parts of human consciousness work? (Why do we want to speculate about seething instincts, drives, emotions, sexual desires, polymorphous perversions, wanting to kill our fathers and sleep with our mothers?) Should we not be bent on rooting out any influences such forces might have (if they really exist at all), including a scientific interest in them? Should we not be banishing the Devil so far away that he can have no influence on our consciousness, let alone deviously work through our *un-consciousness*?

Well no. Not if we wish to have a non-teleological social science and to understand human nature as it really is. So, I have set out to present a model of human nature which suggests that the *how* of consciousness can be conceptualized as the Devil's chaplain (and his psychoanalyst) suggested it would have to be. And, I have argued that, just as there was in the chaplain's message, there is the potential for discovering considerable 'order in process' in this version of human nature - in pre-knowledge, in 'natural patterns' or natural kinds of behaviour, in politics. I have also tried to answer bits of the big question, 'why is human nature like it is ?', and to show some of the 'order in process' of how it translates into political society.

This has been based on current evidence from the neural sciences, human phylogenetic history, psychology and modern developments in psychobiology and sociobiology (and to a lesser degree, anthropology and sociology). The search has been to see why we might have the emotional-cognitive processes we have, and why we have some of the specific species typical desires and fears we have. But much more in this vein must be done. It is one thing to understand something about how human nature might work in a non-teleological sense (and even some of its characteristic motivational impact), but quite another to identify sufficient universal desires and fears which, along with emotional-cognitive motivators, underpin specific patterns of social behaviour. This will have to be the next step in our search for a non-teleological social science.

In the meantime we will have to content ourselves with the observation that rationality has not destroyed reason; reason, in the form of realism, and through science, searching for understanding, lives on. In fact, it never died; Darwinism somehow withstood the adulation of social Darwinists and the criticisms of creationists and social constructionists. It has generated a massive core of dedicated researchers and theoreticians who have not been taken in by human hopes for utopia, perfected selves, womb-like communities or societies equal and fair in all things under the sun. And Freud lives on. Few individuals in human history have had so much of their thought become common currency while being declared to be wrong in almost every aspect.

We will undoubtedly have access to exciting new insights into the characteristics of human nature as they will be uncovered in the next few years of spectacular developments in brain science, evolutionary theory, paleoanthropology, psychobiology, sociobiology and the study of politics *as* political activities (rather than government or political structures). This last may prove to be especially significant for a social science, because it is likely to be the case that what consciousness is for the evolutionary psychologist, politics is for the social scientist. But the search for the Devil's ambassador probably will have to wait upon further findings concerning relatively specific propensities and natural kinds of human behaviour; the work of the Devil's chaplain and of his psychoanalyst is not yet finished.

NOTES

1. Cf., Saint -Simon, 1964; Comte, 1974; Smith, 1981; Durkheim, 1964, 1965, 1982; Marx, 1867; Marx and Engels, 1848, 1951; Spencer, 1862, 1876, 1891; Ward, 1893; Dobb, 1937; Schumpeter, 1943; Rostow, 1971; Rothschild, 1971; Kroeber, 1917, 1948; Mead, M, 1943; Evans-Pritchard, 1937, 1962; Radcliffe-Brown, 1952; White, 1949; Parsons, 1937, 1967; MacIver, 1947; Merton, 1968; Mills, 1959; Habermas, 1976, 1984-87, 1982; Kelly, 1955; Goffman, 1968, 1971; Harre, 1988. On notions of **Progress** see, Alexander and Gill, 1984; Baczko, 1989; Kumar, 1991; Nisbet, 1969; Hertzler, 1965; Manuel, 1973.

2. Cf., Woodburn, 1982; Power, 1988; Boehm, 1997; papers in Lee and DeVore, 1968; Birdsell, 1968; Ambrose, 1975; Service, 1963; Larson, 1997; Turnbull, 1965, 1976; Kluckhohn and Leighton, 1946; Goody, 1983; Harris, 1968; Schneider, 1968 , Marsh, 1967; Lewis, 1968a; Hiatt, 1962; Murdock, 1949.

3. Emotional-Cognitive Processing

Important references for emotional-cognitive processes which emphasize the biological dimension include: Bloom and Lazerson, 1988; papers in *Scientific American,* 'Mind and Brain' (Sept. 1992); Greenfield, 1995; Springer and Deutsch, 1981; Hubel, 1978; Geschwind, 1979; Mahendra, 1987; Panksepp, 1982, 1990; Plutchik, 1980, 1984; Buck, 1984; Cohen, 1979; Hamburg, 1975; Izard, 1982, 1991; Popper and Eccles, 1977; Leventhal, 1984; Gilbert, 1984, 1989; Ornstein, 1973, 1991; Panksepp, Siviy and Normansell, 1985; Mountcastle, 1975, 1978; Hobson, 1990; papers in *Cognition and Emotion* Vol. 1:1 (1987) and, Vol. 4:3 (1990); Freud, 1891, 1940, 1976, Chap VII

Key works specifically concerned with emotions and 'emotional processes' as serious dimension of human motivational processes, but not included above or below are: Freud, 1957, 1961, 1962, 1971, 1976; Hamburg, 1963; Leventhal and Scherer, 1987; Mandler, 1975, 1982; Schneirla, 1957, 1966; Aronson, *et. al.,* 1970; Badcock, 1986, 1994; Buss, A., 1980; Buss, D., 1992, 1994; Planalp, 1999; Darwin, 1872; Klein, 1952; Knapp, 1963; Levi, 1975; Lopreato, 1984; Maslow, 1968, 1987; Rasmussen, 1968, 1968a; Scherer and Ekman, 1984; Strongman, 1973; Arnold, 1960; Wilson and Daly, 1992; Taylor and Brown, 1988; Nesse and Lloyd, 1992.

Major works which primarily focus on advanced cognition but not listed above include: Glass and Holyoak, 1986; Campbell, 1984; Chance and Omark, 1988; Chomsky, 1957, 1972; Lumsden and Wilson, 1985; Crook, 1980; Kelly, 1955; Levi-Strauss, 1966, 1963; Mandler, 1985; Piattelli-Palmarini, 1980; Forgas, 1981; Benjafield, 1992; Eysenck and Keane, 1990; Guidano and Liotti, 1983; Byrne and Whiten, 1988; Whiten and Byrne, 1997; Dunbar, 1996; Barkow, Cosmides and Tooby, 1992; Pinker, 1998

4. Some important references concerning sexual selection and human evolution include: Trivers, 1971, 1972, 1981; Symons, 1979; Hrdy, 1981; Fedigan, 1986; Haraway, 1989; Altmann, 1980; Brin, 1995; Ridley, 1993; Buss, 1992, 1994; Ellis, 1992; Wilson and Daly, 1992 Lovejoy, 1980, 1981; Badcock, 1986, 1991; Miller, 1997; Goy and McEwen, 1980; Alexander, 1971, 1974, 1975, 1979a; Barash, 1981 Morris, 1968; Eibl-Eibesfeldt, 1971; Mellen, 1981; Gould, 1977; Zihlman, 1981, 1983; Tanner, 1981; Dahlberg, 1981; Turke and Betzig, 1985; Hill, 1984; Crook, 1980; Whiten, 1997; Chance, 1988, 1988a.

5. Cf., Machiavelli, 1950; Michels, 1962; Evennett, 1968; Brown, 1969; Kramer, 1969; Walter, 1969; Halberstam, 1969; Sedman, 1970; Janis, 1972; Halperin, 1974; Haldeman, 1978; Stockman, 1986; Zald, 1970; Skidmore, 1967; Sloan, 1972; Galbraith, 1969, 1975; Conquest, 1971; Nove, 1972; Khrushchev, 1977; Stepan, 1971.

REFERENCE

Adorno, T. *et al*, 1969, *The Authoritarian Personality,* N. Y.: Norton.
Alexander, P., and Gill, R., (eds.), 1984, *Utopias,* London: Duckworth.
Alexander, R., 1971, 'The Search for an Evolutionary Philosophy of Man', *Proceedings of the Royal Society of Victoria,* 84: 99-120.
Alexander, R., 1974, 'The Evolution of Social Behavior', *Annual Review of Ecology and Systematics,* 5, pp., 325-383.
Alexander, R., 1975, 'The Search for a General Theory of Behavior', *Behavioral Science,* 20:77-100. .
Alexander, R., 1979, *Darwinism and Human Affairs,* Seattle: U. of Washington Press.
Alexander, R., 1979a, 'Evolution and Culture', In Chagnon and Irons, 1979.
Alexander, R., 1979b, 'Natural Selection and Social Exchange', in Burgess and Hudson, 1979.
Allen, E. *et al*., 1975, 'Against Sociobiology', *New York Review of Books.* (Nov. 13).
Altmann, J., 1980, *Baboon Mothers and Infants,* Camb. Mass.: Harvard U. Press.
Ambrose, S., 1975, *Crazy Horse and Custer: The Parallel Lives of Two American Warriors,* New York: Meridian.
Anglo, S., 1977, *The Damned Art: Essays in the Literature of Witchcraft,* London: Routledge and Kegan Paul.
Appleman, P., (ed.), 1979, *Darwin,* New York: Norton.
Aristotle, 1885, *The Politics,* (Trans. by Benjamin Jowett), Oxford: Clarendon Press.
Arnold, M., 1960, *Emotions and Personality,* New York: Columbia U. Press.
Aronson, L., Tobach, E., Lehrman, D., and Rosenblatt, J., 1970, *Development and Evolution of Behavior: Essays in Memory of T. C.. Schneirla,* San Francisco: W.H. Freeman.
Ayala, F., 1978, 'The Mechanisms of Evolution', *Scientific American* (Vol. 239:3, pp, 56-69 - Sept.).
Baczko, B., 1989, *Utopian Lights,* N.Y.: Paragon.
Badcock, C., 1986, *The Problem of Altruism,* Oxford: Basil Blackwell.
Badcock, C., 1991,` *Evolution and Individual Behavior: An Introduction to Human Sociobiology,* Oxford: Blackwell.
Badcock, C., 1994, *PsychoDarwinism: The New Synthesis of Darwin and Freud,* London: HarperCollins.
Banton, M., (ed.), 1968, *The Relevance of Models for Social Anthropology,* London: Tavistock.
Barash, D., 1981, *Sociobiology: The Whisperings Within,* Glasgow: Fontana.
Barash, D., 1982, *Sociobiology and Behaviour,* London Hodder and Stoughton.
Barkow, J., Cosmides, L. and Tooby, J., 1992, *The Adapted Mind: Evolutionary Psychology and the Generation of Culture,* Oxford: Oxford U. Press.
Barnett, S., 1983, 'Humanity and Natural Selection', *Ethology and Sociobiology,* 4:1, pp. 35-51.
Baron-Cohen, X., 1995, *Mindblindness,* Camb. Mass.: MIT Press.
Barton, A and Dunbar, R., 1997, 'Evolution of the Social Brain', in Whiten and Byrne, 1997.
Baudrillard, J. 1983, *Simulations,* New York: Semiotext(e).
Bauman, Z., 1976, *Towards a Critical Sociology: An Essay on Commonsense and Emancipation,* London: RKP.
Bauman, Z., 1988, "Sociology After the Holocaust", *Br. Jr. of Sociology,* Vol. XXXIX.
Bauman, Z., 1989, *Modernity and the Holocaust,* Camb.: Polity Press.
Baumeister, R., 1997, *Evil: Inside Human Violence and Cruelty,* N.Y.: W. H. Freeman.
Bellamy, E., 1888, *Looking Backward, 2000-1887,* Camb. Mass.: Belknap Press.
Belle, D. (ed.), 1982, *Lives in Stress: Women and Depression,* London: Sage.
Benedict, R., 1946, *Patterns of Culture,* N.Y.: Mentor Books.
Benjafield, J., 1992, *Cognition,* Englewood Cliffs, N.J.: Prentice-Hall.
Berman, M., 1982, *All That is Solid Melts Into Air,* N.Y.: Simon and Schuster.
Best, S., and Kellner, D., 1991, *Postmodern Theory: Critical Interrogations,* London: Macmillan.

Bhaskar, R., 1986, *Scientific Realism and Human Emancipation,* London: Verso.
Birdsell, J., 1968, 'Some Predictions from the Pleistocene Based on Equilibrium Systems among Recent Hunter-Gatherers', in Lee and DeVore, 1968.
Blackmore, S., 1999, *The Meme Machine,* Oxford: Oxford U. Press.
Blau, P. and Duncan, O., 1967, *The American Occupational Structure,* N.Y.: John Wiley.
Bloom, F. and Lazerson, A., 1988, *Brain, Mind, and Behavior,* N.Y.: Freeman and Co.
Boas, F., 1965, *The Mind of Primitive Man,* N.Y.: The Free Press.
Boas, F., 1966, *Race, Language and Culture,* N.Y.: The Free Press..
Boehm, C., 1997, 'Egalitarian Behaviour and the Evolution of Political Intelligence', in Whiten and Byrne, 1997.
Bowler, P., 1989, *The Invention of Progress,* Oxford: Basil Blackwell.
Brin, D., 1995, 'Neoteny and Two-Way Sexual Selection in Human Evolution: Paleo-Anthropological Speculation on the Origins of Secondary-Sexual Traits, Male Nurturing and the Child as a Sexual Image', at, http://www.kithrup.com/brin/neoteny.html
Brogan, H., 1986, *The Pelican History of the United States of America,* Harmondsworth: Penguin.
Brown, A., 2000, *The Darwin Wars,* London: Touchstone.
Brown, G. and Harris, T., 1978, *Social Origins of Depression,* N.Y.: Free Press.
Brown, Henry Phelps, 1977, *The Inequalities of Pay,* London: Oxford U. Press.
Brown, R., 1969, 'The American Vigilante Tradition', in Graham and Gurr, 1969.
Buck, R, 1984, *The Communication of Emotions,* London: The Guilford Press.
Burgess, R. and Huston, T. (eds), *Social exchange in Developing Relationships,* N.Y.: Academic Press.
Buss, A., 1980, *Self-Consciousness and Social Anxiety,* San Francisco: W.H. Freeman.
Buss, D., 1992, 'Mate Preference Mechanisms: consequences for Partner Choice and Intersexual Competition', in Barkow, Cosmides and Tooby, 1992.
Buss, D., 1994, *The Evolution of Desire,* N.Y.: Basic Books.
Byrne, W. and Whiten, A., (eds), 1988, *Machiavellian Intelligence,* Oxford: Clarendon Press.
Callinicos, A., 1985, 'Postmodernism, Post-Structuralism, Post Marxism ?' *Theory, Culture and Society,* Vol. 2:3, pp, 85-102.
Callinicos, A., 1990, *Against Postmodernism: A Marxist Critique,* N.Y.: St. Martin's Press.
Campbell, B., (ed), 1972, *Sexual Selection and the Descent of Man,* Chicago: Aldine.
Campbell, J., 1984, *Grammatical Man,* Harmondsworth: Penguin.
Carroll, J., 1985, *Guilt,* London: Routledge and Kegan Paul.
Cawelti, J., 1965, *Apostles of the Self-Made Man: Changing Concepts of Success in America,* Chicago: U. of Chicago Press.
Chagnon, N. and Irons, W., (eds), 1979, *Evolutionary Biology and Human Social Behavior,* North Scituate, Mass.: Duxbury Press.
Chagnon, N., 1979, 'Is Reproductive success "equal" in Egalitarian Societies?', in Chagnon and Irons, 1979.
Chance, M., 1988, 'Introduction', in Chance and Omark, 1988.
Chance, M., 1988a, 'A Systems Synthesis of Mentality', in Chance and Omark, 1988.
Chance, M., Omark, D., (eds.), 1988, *Social Fabrics of the Mind,* London: Lawrence Erlbaum Associates.
Changeux, J-P., 1980, 'Properties of the Neural Network', in Piattelli-Palmarini, 1980.
Chodorow, N., 1978, *The Reproduction of Mothering: Psychoanalysis and the Sociology of Gender,* Berkeley: U. of California Press.
Chodorow, N., 1990, *Feminism and Psychoanalytic Theory,* New Haven: Yale U. Press.
Chomsky, N., 1957, *Syntactic Structures,* The Hague: Mouton and Co.
Chomsky, N., 1972, *Language and Mind,* N.Y.: Harcourt Brace Jovanovich.
Clark, M. and Leiter, B., 1997, *Nietzsche: Daybreak; Thoughts on the Prejudices of Morality,* Cambridge: Camb. U. Press.
Clark, M. and Leiter, B., 1997a, 'Introduction', Clark, M. and Leiter, B., 1997.
Clegg, S., 1989, *Frameworks of Power,* Beverly Hills: Sage.

Clutton-Brock, T. and Harvey, P. (eds.), *Readings in Sociobiology*, San Francisco: W. H. Freeman.

Cohen, D., 1979, *Sleep and Dreaming: Origins, Nature and Functions,* N.Y.: Pergamon.

Cohn, N., 1970, *The Pursuit of the Millennium,* London: Paladin.

Comte, A., 1974, *The Positive Philosophy,* New York: AMS Press.

Conquest, R., 1971, *The Great Terror,* Harmondsworth: Penguin.

Conway, D., 1996, *Nietzsche and the Political,* London: Routledge.

Cooley. C. 1902, ` *Human Nature and the Social Other,* New York: Charles Scribners.

Cosmides, L., Tooby, J. and Barkow, H., 1992, 'Introduction: Evolutionary Psychology and Conceptual Integration', in Barkow, Cosmides, and Tooby, 1992.

Craib, I. 1984, *Modern Social Theory,* New York: Harvester Wheatsheaf.

Crick, F., 1979, 'Thinking about the Brain', *Scientific American,* Vol. 241:3 (Sept.).

Cronin, H., 1991, *The Ant and the Peacock,* Camb.: Camb. U. Press.

Cronk, L., Chagnon, N. and Irons, W. (eds), 2000, *Adaptation and Human Behavior,* N.Y.: Aldine De Gruyter.

Crook, J., 1980, *The Evolution of Human Consciousness,* London: Oxford University Press.

Csikszentmihalyi, M., 1975, *Beyond Boredom and Anxiety,* San Francisco: Jossey Bass.

Dahlberg, F., 1981, *Woman the Gatherer,* New Haven: Yale U. Press.

Damasio, A. and Damasio, H., 1992, 'Brain and Language', *Scientific American,* Vol. 267:3 (Sept.).

D'Amico, R., 1986, 'Going Relativist', *Telos,* 67.

Darwin, C., 1859 (1996), *On the Origin of Species by Means of Natural Selection, or the Preservation of Favoured Races in the Struggle for Life,* Oxford: Oxford U. Press.

Darwin, C., 1872, *The Expression of the Emotions in Man and Animals,* Chicago: U. of Chicago Press.

Darwin, C., 1874, *The Descent of Man and Selection in Relation to Sex* (second ed.), London: Murray.

Darwin, C., 1950, *Charles Darwin's Autobiography,* (ed. by Francis Darwin), N.Y.: Schuman.

Davies, C., 1982, 'Sexual Taboos and Social Boundaries', *Amer. Jr. of Sociology,* Vo. 87 (March).

Davies, C., 1990, *Ethnic Humour Around the World: a Comparative Analysis,* Bloomington: Indiana U. Press.

Davies, C., 1990a, 'Review' of *Humor and Society: Explorations in the Sociology of Humor,* in *Reviewing Sociology,* Vol. 7: 1, pp., 28-31.

Davies, J., 1969, 'The J-Curve of Rising and Declining Satisfactions as a Cause of Some Great Revolutions and a Contained Rebellion', in Graham and Gurr, 1969.

Dawkins, R., 1976, *The Selfish Gene,* London: Oxford U. Press.

Dawkins, R., 1982, *The Extended Phenotype,* N.Y.: Oxford U. Press.

Dawkins, R., 1986, *The Blind Watchmaker,* London: Longman.

Dawkins, R., 1989, *The Selfish Gene* (New Edition), Oxford: Oxford U. Press.

Dawkins, R., 1997, *Climbing Mount Improbable,* Harmondsworth: Penguin.

Dennett, D., 1995, *Darwin's Dangerous Idea,* London: Allen Lane.

Desmond, A. and Moore, J., 1991, *Darwin,* London: Michael Joseph.

Desmond, A., 1989, *The Politics of Evolution,* Chicago: U. of Chicago Press.

Dinnerstein, D., 1976, *The Mermaid and the Minotaur,* New York: Harper and Row.

Dobb, M., 1937, *Political Economy and Capitalism,* London: Routledge and Kegan Paul.

Dollard, J., 1937, *Caste and Class in a Southern Town,* New Haven: Yale U. Press.

Douglas, M. (ed.), 1970a, *Witchcraft Confessions and Accusations,* London: Tavistock.

Douglas, M., 1970, *Purity and Danger,* Harmondsworth, Penguin.

Douglas, M., 1975, *Implicit Meanings,* London: Routledge.

Dunbar, R., (ed.)., 1993, *Human Reproductive Decisions: Biological and Social Perspectives,* Galton Institute symposium, 13th), London: Macmillan.

Dunbar, R., 1996, *Grooming, Gossip and the Evolution of Language,* London: Faber and Faber.

Durkheim, E., 1964, *The Division of Labor in Society,* N.Y.: The Free Press.

Durkheim, E., 1965, *The Elementary Forms of the Religious Life,* N.Y.: The Free Press.
Durkheim, E., 1982, *The Rules of The Sociological Method,* London: Macmillan.
Edelman, G. and Mountcastle, V., 1978, *The Mindful Brain,* Camb. Mass.: MIT Press.
Eibl-Eibesfeldt, I., 1971, *Love and Hate,* London: Methuen.
Eiseley, L., 1958, *Darwin's Century,* New York: Doubleday.
Ekman, P. and Oster. H., 1979, 'Facial Expressions of Emotion, *Annual Rev. of Psychology,* 30.
Ellis, B., 1992, 'The Evolution of Sexual Attraction: Evaluative Mechanisms in Women', in Barkow *et. al.*, 1992.
Epstein, S., 1992, 'Gay Politics, Ethnic Identity: the Limits of Social Constructionism', in Stein, 1992.
Erikson, E., 1965, *Childhood and Society,* N.Y.: Vintage Books.
Evans-Pritchard, E., 1937, *Witchcraft, Oracles and Magic Among the Azande,* Oxford: Clarendon Press.
Evans-Pritchard, E., 1962, *Essays in Social Anthropology,* London: Faber and Faber.
Evennett, H., 1968, *The Spirit of the Counter-Reformation,* Camb.: Camb. U. Press.
Eysenck, M. and Keane, M., 1990, *Cognitive Psychology,* Lawrence Erlbaum Associates.
Fedigan, L., 1986, 'The Changing Role of Women in Models of Human Evolution', *Annual Rev. of Anthropology,* 15.
Ferguson, A., 1966, *An Essay on the History of Civil Society,* Edinburgh: Edinburgh U. Press.
Fiedler, L., 1981, *Freaks: Myths and Images of the Secret Self,* Harmondsworth: Penguin.
Fischbach, G., 1992, 'Mind and Brain', *Scientific American,* Vol. 267:3 (Sept.).
Fitzgerald, M., 1978, *Human Embryology,* London: Harper and Row.
Fitzgerald, R., (ed.), 1977, *Human Needs and Politics,* Oxford: Pergamon.
Fitzgerald, R., 1977a, 'The Ambiguity and Rhetoric of "Need"', in Fitzgerald, 1977.
Foley, R., 1996, 'An Evolutionary and Chronological Framework for Human Behaviour', in Runciman, Maynard Smith and Dunbar, 1996.
Forgas, J., (ed.), 1981, *Social Cognition,* London: Academic Press.
Forgas, J., 1981a, 'Preface', in Forgas, 1981.
Forgas, J., 1981b 'Epilogue: Everyday Understandings and Social Cognition' in Forgas, 1981.
Fortey, R., 1998, *Life: An Unauthorised Biography,* London: Flamingo.
Foster, G., 1972, 'The Anatomy of Envy: A Study in Symbolic Behavior', *Current Anthropology,* 13: pp., 165-202.
Foucault, M. 1988, *Politics, Philosophy, Culture: Interviews and other Writings 1977-1984,* N.Y.: Routledge.
Foucault, M., 1979, *Discipline and Punish,* N.Y.: Vintage Books.
Foucault, M., 1980, *Power/Knowledge: Selected Interviews and Other Writings, 1972-1977,* N.Y.: Pantheon.
Foucault, M., 1980a, *The History of Western Sexuality: An Introduction,* Vol. 1, N.Y.: Vintage Books.
Fox, R., 1967, *Kinship and Marriage: An Anthropological Perspective* Harmondsworth: Penguin Books.
Fox, R., 1980, *The Red Lamp of Incest,* London: Hutchinson.
Fox, R., 1986, 'Fitness by Any Other Name', *Behavior and Brain Science,* 9:192-93.
Freeman, D., 1984, *Margaret Mead and Samoa: The Making and Unmaking of an Anthropological Myth,* Harmondsworth: Penguin.
Freud, A., 1966, *The Ego and the Mechanisms of Defense,* N.Y.: International U. Press.
Freud, S., 1891 (1953) *On Aphasia: A Critical Study,* London: Imago.
Freud, S., 1895 (1954), 'Project for a Scientific Psychology' *The Origins of Psychoanalysis, Letters to Wilhelm Fliess, Drafts and Notes: 1887-1902,* London: Imago.
Freud, S., 1905, *Jokes and their Relationship to the Unconscious,* in Freud, 1974.
Freud, S., 1940, 'An Outline of Psycho-Analysis', *in Freud, 1974.*
Freud, S., 1957, *Civilization and Its Discontents,* London: The Hogarth Press.
Freud, S., 1961 (1927), *The Future of an Illusion,* N.Y.: Norton.
Freud, S., 1962, *Three Essays on The Theory of Sexuality,* N.Y.: Basic Books.

Freud, S., 1971, (1916-1917), *Complete Introductory Lecturers on Psychoanalysis,* London: George Allen and Unwin.
Freud, S., 1974, *The Standard Edition of the Complete Works of Sigmund Freud,* Vol 23, (trans. James Strachey), London: Hogarth.
Freud, S., 1976 (1930), *The Interpretation of Dreams,* Trans., James Strachey, London: Penguin.
Friday, N., 1980, *Men in Love,* London: Arrow Books.
Fromm, E., 1960, *Escape From Freedom,* London: Routledge and Kegan Paul.
Galbraith, J K., 1975, *Economics, Peace and Laughter,* Harmondsworth: Penguin.
Galbraith, J. K., 1969, *The New Industrial State,* Harmondsworth: Penguin.
Garfield, P., 1976, *Creative Dreaming,* N.Y.: Ballantine Books.
Gay, P., 1954, 'The Enlightenment in the History of Political Thought', *Political Science Quarterly,* LXIX, (Sept.).
Gay, P., 1967, *The Enlightenment: An Interpretation, The Rise of Modern Paganism,* (Vol. 1) London: Weidenfeld and Nicolson.
Gay, P., 1969 *The Enlightenment: an Interpretation.: The Science of Freedom,* (Vol. 2) London: Weidenfeld and Nicolson.
Gay, P., 1988, *Voltaire's Politics: The Poet as Realist,* New Haven: Yale U. Press.
Gay, P., 1990, *Reading Freud: Explorations and Entertainments,* New Haven: Yale U. Press.
Gay, P., 1993, *The Cultivation of Hatred,* N.Y.: W. W. Norton.
Gershon, E. and Rieder, R., 1992, 'Major Disorders of Mind and Brain', *Scientific American,* Vol 267:3 (Sept.).
Geschwind, N., 1979, 'Specializations of the Brain', *Scientific American,* offprint 241.
Gibson, J., 1979, *The Ecological Approach to Visual Perception,* Boston: Houghton Mifflin.
Giddens, A., 1985, *The Nation State and Violence,* Camb.: Polity.
Gilbert, P., 1984, *Depression: From Psychology to Brain State',* London: Lawrence Erlbaum.
Gilbert, P., 1989, *Human Nature and Suffering,* Hove and London: Lawrence Erlbaum.
Girard, R., 1978, *To Double Business Bound: Essay on Literature, Mimesis and Anthropology,* Baltimore: Johns Hopkins Press.
Glass, A. and Holyoak, K., 1986, *Cognition,* N.Y.: Random House.
Goffman, E., 1968, *Asylums,* Harmondsworth: Penguin.
Goffman, E., 1971, *The Presentation of Self in Everyday Life,* Harmondsworth: Penguin.
Goldsmith, T., 1991, *The Biological Roots of Human Nature,* Oxford: Oxford U. Press.
Goldstein, A. and Michaels, G., 1985, *Empathy: Development, Training and Consequences,* Hillsdale N.J.: Lawrence Erlbaum.
Goldthorpe, J. *et. al.*, 1980, *Social Mobility and Class Structure in Modern Britain,* London: Oxford U. Press.
Goody, J., 1983, *The Development of the Family and Marriage in Europe,* Camb.: Camb. U. Press.
Gould, S., 1977, *Ontogeny and Phylogeny,* Camb. Mass.: Harvard U. Press.
Gould, S., 1996, *Dinosaur in a Haystack,* London: Jonathan Cape.
Gould, S., 1996a, 'Can We Complete Darwin's Revolution', in Gould, 1996.
Gould, S., 1997, 'Darwinian Fundamentalism', *The New York Review of Books,* (June 12, 1997).
Goy, R., and McEwen, B. (eds.), 1980, *Sexual Differentiation of the Brain,* Camb. Mass.: M.I.T. Press.
Graham, H. and Gurr, T. (eds.), 1969, *Violence in America,* N.Y.: Signet, The New American Library.
Granovetter, M., 1979, 'The Idea of Advancement in Theories of Social Evolution and Development', *AJS,* Vol 85, No. 3.
Gray, J., 2000, 'Twenty Years of Evolutionary Biology and Human Social Behavior: Where are We Now?, in Cronk, Chagnon and Irons, 2000.
Green, C. and McCreery, C., 1994, *Lucid Dreaming: The Paradox of Consciousness During Sleep,* London: Routledge.
Greenfield, S., 1995, *Journey to the Centre of the Mind,* New York: Freeman and Co.

Grove, W., 1970, 'Societal Reactions as an Explanation of Mental Illness: An Evaluation', *Amer. Soc. Review,* 35 (Oct.).

Guidano, V. and Liotti, G., 1983, *Processes and Emotional Disorders,* London: The Guilford Press.

Gurr, T., 1970, *Why Men Rebel,* New Brunswick: Princeton U. Press.

Habermas, J., 1971, *Knowledge and Human Interest,* Boston: Beacon Press.

Habermas, J., 1976, *Legitimation Crisis,* London: Heinemann.

Habermas, J., 1984 and 1987, (Vols. 1 and 2), *Theory of Communicative Action,* Boston: Beacon Press.

Halberstam, D., 1969, *The Best and the Brightest,* Greenwich, Conn.: A Fawcett Crest Book.

Haldeman, H. R., 1978, *The Ends of Power,* London: A Star Book, W.H. Allen.

Halperin, M., 1974, *Bureaucratic Politics and Foreign Policy,* Washington D.C.: The Brookings Institute.

Hamburg, D., 1963, 'Emotions in the perspective of Human Evolution', in, Knapp, 1963.

Hamburg, D., *et. al.,* 1975, 'Anger and Depression in Perspective of Behavioral Biology', in Levi, 1975.

Hamilton, M. 1989, Review of: *Beliefs and Ideology* (Ellis Horwood/Tavistock, 1986) by Kenneth Thompson, *Reviewing Sociology,* Vol 6:2.

Hamilton, M., 1987, 'The Elements of the Concept of Ideology, *Political Studies,* 35:, pp, 18-38.

Hamilton, M., 1995, *The Sociology of Religion,* London: Routledge.

Hamilton, M., and Hirszowicz, M., 1993, *Class and Inequality: Comparative Perspectives,* Hemel Hempstead: Harvester.

Hamilton, W., 1963, 'The Evolution of Altruistic Behaviour', in Clutton-Brock, 1978.

Hamilton, W., 1964, 'The Genetical Evolution of Social Behaviour', *Jr. of Theoretical Biology,* 7: 1-52.

Hamilton, W., 1971, 'Geometry for the Selfish Herd', *Jr. of Theoretical Biology,* 31:295-311.

Handlin, O., 1951, *The Uprooted: The Epic Story of the Great Migrations that Made the American People,* Boston: Little Brown.

Handlin, O., 1966, *The American People,* Camb. Mass.: Harvard U. Press.

Haraway, D., 1989, *Primate Visions: Gender, Race, and Nature in the World of Modern Science,* London: Routledge.

Harre, R. (ed.), 1988, *The Social Construction of Emotions,* Oxford: Basil Blackwell.

Harris, D. (ed.), 1957, *The Concepet of Development,* Minneapolis: U. of Minnesota Press.

Harris, H., 1980, *The Principles of Human Biochemical Genetics,* Amsterdam: Elsevier.

Harris, M., 1968, *The Rise of Anthropological Theory,* London: Routledge and Kegan Paul.

Harris, M., 1977, *Cows, Pigs, Wars and Witches,* Glasgow: Fontana.

Harris, M., 1979, *Cultural Materialism: The Struggle for a Science of Culture,* N.Y.: Random House.

Hatab, L., 1995, *A Nietzschean Defense of Democracy,* Chicago: Open Court.

Hauser, M., 1997, 'Minding the Behaviour of Deception', in Whiten and Byrne, 1997.

Hearne, K., 1986, 'Dream Sense', *Nursing Times,* (Jan. 1).

Herbert, M., 1983, 'Evolutionary Theory in Ferment', *Telos,* No. 57.

Hertzler, J., 1965, *The History of Utopian Thought,* N.Y.: Cooper Square Pub.

Hiatt, L., 1962, 'Local Organization among the Australian Aborigines, *Oceania,* 32: 267-86.

Hicks, G. and Leis, P. (eds.), 1977, *Ethnic Encounters: Identities and Contexts,* North Scituate, Mass.: Duxbury.

Hill, J., 1984, 'Human Altruism and Sociocultural Fitness', *Jr. of Soc. and Boil. Structures,* 7:17-35.

Hinde, R. and Stevensen-Hinde, J., 1973, *Constraints on Learning,* N.Y.: Academic Press.

Hinde, R., 1982, *Ethology,* Glasgow: Fontana.

Hirst, P., 1976, *Social Evolution and Sociological Categories,* London: Allen and Unwin.

Ho. D., 1976, 'On the Concept of Face', *Am. J. of Soc.* 81.

Hobbes, T., 1968, *Leviathan,* (C. Macpherson ed.), Harmondsworth: Penguin.

Hobson, A., 1990, *The Dreaming Brain,* London: Penguin.
Hofstadter, R., 1955, *Social Darwinism in American Thought,* Boston: Beacon.
Hollingdale, R., 1973, *Nietzsche,* London: Routledge and Kegan Paul.
Hopkins, K., 1968, 'Structural Differentiation in Rome (200 - 31 B.C.): The Genesis of an Historical Bureaucratic Society', In Lewis, 1968.
Horkheimer, M., and Adorno, T., 1972, *Dialectic of Enlightenment,* New York: Seabury.
Hrdy, S., 1981, *The Woman that never Evolved,* Camb. Mass: Harvard U. Press.
Hubel, D. (ed.), 1978, *The Brain,* N.Y.: Freeman.
Hughes, S., 1959, *Consciousness and Society: The Reorientation of European Social Thought, 1890-1930,* London: MacGibbon & Kee.
Hull, D., 1978, 'The Sociology of Sociobiology', *New Scientist,* 79.
Hull, D., 1978a, 'The Sociobiology Debate', *New Scientist,* 21 (Sept.).
Hume, D., 1963, *Essays: Moral, Political and Literary,* Oxford: Oxford U. Press.
Hume, D., 1964 (1882), *Philosophical Works,* Green, T. and Grose, T., Darmstadt: Scientia Verlag Aalen.
Hume, D., 1964a (1882), 'The Sceptic', in Hume, 1964.
Hume, D., 1968 (1888), *A Treatise of Human Nature,* Oxford: Clarendon Press.
Hume, D., 1975, *Enquires Concerning Human Understanding and Concerning the Principles of Morals,* Oxford: Oxford U. Press.
Humphrey, N., 1978, 'Nature's Psychologists', *New Scientist,* 79.
Hutcheon, L,, 1989, *The Politics of Postmodernism,* London: Routledge.
Inkeles, A., and Rossi, P., 1956, "National Comparisons of Occupational Prestige", *Am. Jr. of Sociology,* 61.
Irons and Cronk, 2000, 'Two Decades of a New Paradigm', in Cronk, Chagnon and Irons, 2000.
Irons, W., 1979, 'Natural Selection, Adaptation and Human Social Behaviour', in Chagnon and Irons, 1979.
Izard, C. (ed.), 1982, *Measuring Emotions in Infants and Children,* Camb.: Camb. U. Press.
Izard, C., 1991, *The Psychology of Emotions,* N.Y.: Plenum.
Janis, I., 1972, *Victims of Groupthink,* Camb. Mass.: Houghton Mifflin.
Jarvie, I. C., 1984, *Rationality and Relativism: In Search of a Philosophy and History of Anthropology,* London: Routledge and Kegan Paul.
Johanson, D. and Edey, M., 1981, *Lucy: The Beginnings of Humankind,* London: Granada.
Jolly, A., 1988, 'The Evolution of Purpose', in Byrne and Whiten, 1988.
Jones, S., 1999, *Almost Like a Whale,* London: Doubleday.
Jouvet, M., 1973, 'Commentary' in Webb, 1973.
Jouvet, M., 1974 'The Role of Monoaminergic Neurons in the Regulation and Function of Sleep', in Petre-Quadens and Schlag, 1974.
Jung, C., 1972, *Four Archetypes,* London: Routledge and Kegan Paul.
Jung, C., 1984, *Modern Man in Search of a Soul,* London: Ark Paperbacks.
Kandel, E. and Hawkins, R., 1992, 'The Biological Basis of Learning and Individuality', *Scientific American,* Vol. 267:3 (Sept.).
Kant, I., 1963, *Idea for a Universal History,* ed. by Lewis Beck, Indianapolis: Bobbs-Merrill.
Kant, I., 1974, *Groundwork for the Metaphysic of Morals,* (H.J. Paton Trans.), London: Hutchinson.
Kaufmann, W., 1954, *The Portable Nietzsche,* N.Y.: Viking Press.
Kelly, G., 1955, *The Psychology of Personal Constructs,* N.Y.: Norton.
Keohane, N., Rosaldo, Z., Gelpi, C., (eds.), 1982 *Feminist Theory: A Critique of Ideology,* Brighton: Harvester.
Kerrigan, J. F., 1996, *Revenge Tragedies: Aeschylus to Armageddon,* Oxford: Oxford U. Press.
Khrushchev, N., 1977, *Khrushchev Remembers,* (Trans. and ed. by S. Talbot), London: Sphere Books.
Kieckhefer, R., 1976, *European Witch Trials,* London: Routledge and Kegan Paul.
Klein, M., 1952, *Some Theoretical Conclusions Regarding the Emotional Life of Infants,* London: Hogarth.

Kluckhohn, C., 1944, *Navaho Witchcraft,* Boston: Beacon Press.
Kluckhohn, C., and Leighton, D., 1946, *The Navaho,* Boston: Beacon Press.
Knapp, P. H., (ed), 1963, *Expressions of Emotions in Man,* New York: International U. Press.
Kramer, H., Sprenger, J., 1928, (1486), *Malleus Maleficarum,* (Trans. by Montague Summers), London: Arrow Books Ltd. (1971).
Kramer, J., 1969, 'Don't Squeal', *New Society,* (Oct. 2).
Kroeber, A., 1917, 'The Superorganic', *American Anthropologist,* 21: 235-263.
Kroeber, A., 1948, *Anthropology,* San Francisco: Harcourt Brace.
Kroeber, A., 1952, *The Nature of Culture,* Chicago: U. of Chicago Press.
Kumar, K., 1978, *Prophecy and Progress,* Harmondsworth: Penguin Books.
Kumar, K., 1991, *Utopianism,* Milton Keynes: Open U. Press.
Kuper, A., 1983, *Anthropology and Anthropologists: The Modern British School,* London: Routledge and Kegan Paul.
La Rochefoucauld, 1959 (1665), *La Rochefoucauld Maxims,* (L. W.. Tancock, Trans.), Harmondsworth: Penguin.
Laing, R., 1965, *The Divided Self,* Harmondsworth: Penguin.
Larson, R., 1997, *Red Cloud: Warrior-Statesman of the Lakota Sioux,* Norman: University of Oklahoma Press.
Lash, S, 1990, *Sociology of Postmodernism,* London: Routledge.
Lea, C., 1957, *Materials Toward a History of Witchcraft,* London: Thomas Yoseloff.
Leach, E., 1954, *Political Systems of Highland Burma: A study of Kachin Social Structure,* London: G. Bell and Sons Ltd.
Leach, E., 1967, 'Genesis as Myth', in Middleton, 1967.
Leach, E., 1982, *Social Anthropology,* Glasgow: Fontana.
Lee, R. and DeVore, I., 1968, *Man the Hunter,* Chicago: Aldine Pub. Co.
Legman, G., 1971, *Rationale of the Dirty Joke: An Analysis of Sexual Humor,* N.Y.: Grove Press.
Lengermann, P. and Niebrugge-Brantley, J., 1992, 'Contemporary Feminist Theory', in Ritzer, 1992.
Leventhal, H. and Scherer, K., 1987, 'The Relationship of Emotions and Cognition: A Functional Approach to a Semantic Controversy', *Cognition and Emotion,* 1:1, 3-28.
Leventhal, H., 1984, 'A Perceptual Motor Theory of Emotions', in Scherer and Ekman (eds.), 1984.
Levi, L., (ed.), 1975, *Emotions: Their Parameters and Measurement,* N.Y.: Raven Press.
Levi-Strauss, C., 1963, *Totemism,* Boston: The Beacon Press.
Levi-Strauss, C., 1966, *The Savage Mind,* London: Weidenfeld and Nicolson.
Levy, O. (ed.), 1964, *The Complete Works of Friedrich Nietzsche,* N.Y.: Russell and Russell.
Lewis, I. (ed.), 1968, *History and Social Anthropology,* London: Tavistock.
Lewis, I., 1968a, 'Problems in the Comparative Study of Unilineal Descent', in Banton, 1968.
Lewontin, R., 1974, *The Genetic Basis for Evolutionary Change,* N.Y.: Columbia U. Press.
Lewontin, R., 1978, 'Adaptation', *Sci. American,* (Sept.).
Lewontin, R., Rose, S. and Kamin, L., 1984, *Not in Our Genes: Biology, Ideology and Human Nature,* N.Y.: Pantheon.
Lipset, S. and Bendix, R., 1959, *Social Mobility in Industrial Societies,* Berkeley: U. of Calif. Press.
Locke, J., 1881, *Locke's Conduct of the Understanding* (Ed., Thomas Fowler), Oxford: Clarendon Press.
Lopreato, J., 1984, *Human Nature and Biocultural Evolution,* London: George Allen and Unwin.
Lovejoy, O., 1980, 'Hominid Origins: the Role of Bipedalism', *American Jr. of Physical Anthro.* 52 (Feb.).
Lovejoy, O., 1981, 'Is it a Matter of Sex', Chap. 16 in Johanson and Edey, 1981.
Lovibond, S., 1989, 'Feminism and Postmodernism', *New Left Review,* no. 178, pp., 5-28.
Lumsden, C. and Wilson, E., 1981, *Genes, Mind, and Culture,* Camb. Mass.: Harvard U. Press.

Lumsden, C. and Wilson, E., 1983, *Promethean Fire,* Camb. Mass.: Harvard U. Press.
Lumsden, C. and Wilson, E., 1985, 'The Relation Between Biological and Cultural Evolution', *Jr. of Social and Biological Structures,* (Vol. 8:4).
Lynd, R. and Lynd, H., 1937, *Middletown In Transition: A Study in Cultural Conflicts,* N.Y.: Harcourt Brace and Co.
Lyotard, J., 1984, *The Post Modern Condition,* Minneapolis: U. of Minnesota Press.
Maasen, S., Mendelsohn, E. and Weingart, P. (eds), 1995, *Biology as Society, Society as Biology: Metaphors,* Dordrecht: Kluwer Academic Press.
Machiavelli, N., 1950, *The Prince and the Discourses,,* N.Y.: Random House.
Machiavelli, N., 1963, *The Prince,* N.Y.: Pocket Books.
MacIver, R., 1947, *The Web of Government,* New York: Macmillan.
MacLean, P., 1978, 'The Evolution of Three Mentalities', in Washburn and McCown, 1978.
Magnusson, M. and Palsson, H. (trans), 1960, *Njal's Saga,* Harmondsworth: Penguin.
Mahendra, B., 1987, *Depression: The Disorder and its Associations,* Lancaster, U.K.: M.T.P. Press.
Mair, L., 1962, *Primitive Government,* Harmondsworth: Penguin Books.
Malleus, 1486, *Malleus Maleficarum,* in Kramer and Sprenger, 1928.
Mandler, G., 1975, *Mind and Emotion,* N.Y.: Wiley.
Mandler, G., 1982, 'The Construction of Emotion in the Child', in Izard, 1982.
Mandler, G., 1985, *Cognitive Psychology: An Essay in Cognitive Science,* Hillsdale, N.J.:, Lawrence Erlbaum.
Mannheim, K., 1960, *Ideology and Utopia,* London: RKP.
Manuel, F., 1973, *Utopias and Utopian Thought,* London: Souvenir Press.
Marcuse, H., 1962, *Eros and Civilization,* New York: Vintage Books.
Marcuse, H., 1964, *One Dimensional Man,* Boston: Beacon Press.
Marsh, R., 1967, *Comparative Sociology,* N.Y.: Harcourt Brace and World.
Marwick, A., 1981, *Class: Images and Reality in Britain, France and the USA since 1930,* Glasgow: Fontana.
Marwick, M. (ed.), 1970, *Witchcraft and Sorcery,* Harmondsworth: Penguin Books.
Marx K. and Engels, F., 1848, *Manifesto of the Communist Party,* N.Y.: International Pub.
Marx, K. and Engels, F., 1951, *Selected Works,* London: Lawrence and Wishart.
Marx, K., 1867, *Capital: A Critique of Political Economy,* Vol. 1, N.Y.: International Pub.
Maslow, A., 1968, *Towards a Psychology of Being,* N.Y.: Van Nostrand.
Maslow, A., 1987, *Motivation and Personality,* New York: Harper and Row.
Masters, W. and Johnson, V., 1970, *Human Sexual Inadequacy,* Boston: Little. Brown and Co.
Mayr, E., 1961, 'Cause and Effect in Biology', *Science,* 134:1501-6.
Mayr, E., 1978, 'Evolution', *Scientific American,* (Sept.).
Mead, G., 1962, *Mind, Self and Society: From the Standpoint of a Social Behaviorist,* Chicago: U. of Chicago Press.
Mead, M., 1943, *Coming of Age in Samoa,* Harmondsworth: Penguin.
Mellen, S., 1981, *The Evolution of Love,* San Francisco: W.H. Freeman
Merton, R., 1968, *Social Theory and Social Structure,* N.Y.: Free Press.
Michels, R., 1962, *Political Parties,* N.Y.: The Free Press.
Middleton, J. (ed.), 1967, *Myth and Cosmos: Readings in Mythology and Symbolism,* N.Y.: The Natural History Press.
Midelfort, H., 1972, *Witch Hunting in Southwestern Germany,* Palo Alto: Stanford U. Press.
Miller, G., 1997, 'Protean Primate: The Evolution of Adaptive Unpredictability in Competition and Courtship', in Whiten and Byrne, 1997.
Miller, L., 1987, 'REM Sleep: Pilot Light of the Mind', *Psychology Today,* Vol. 21:9 (Sept.), pp, 8-10.
Mills, C. W., 1959, *The Sociological Imagination,* New York: Oxford U. Press.
Mitchell, R., 1975, *Depression,* Harmondsworth: Penguin.
Money, J., 1980, *Love and Lovesickness,* Baltimore: Johns Hopkins U. Press.
Morgenthau, H., 1946, *Scientific Man Versus Power Politics,* Chicago: U. of Chicago Press.

Morgenthau, H., 1954, *Politics Among Nations,* N.Y.: Alfred A. Knopf.

Morris, D., 1968, *The Naked Ape,* London: Corgi Books.

Mountcastle, V., 1975, 'The View from Within: Pathways to the Study of Perception', *Johns Hopkins Medical Journal,* 136

Mountcastle, V., 1978, 'An Organizing Principle for Cerebral functions: The Unit Module and the Distributed System', in Edelman and Mountcastle, 1978.

Murdock, G., 1949, *Social Structure,* N.Y.: Macmillan.

Nesse, R. and Lloyd, A., 1992, 'The Evolution of Psychodynamic Mechanisms', in Barkow, Cosmides and Tooby, 1992.

Nietzsche, F. 1889, *Twilight of the Idles,* in Kaufmann, 1954.

Nietzsche, F., 1886, *Human, All Too Human,* in Levy, 1964.

Nietzsche, F., 1887, *Daybreak,* in Levy, 1964.

Nietzsche, F., 1887a (1967),. *On the Genealogy of Morals,* (Trans. Kaufmann and Hollingdale), N.Y.: Vintage Books.

Nietzsche, F., 1961, *Thus Spoke Zarathustra,* Harmondsworth: Penguin.

Nietzsche, F., 1968, *The Will To Power,* (Trans. W. Kaufmann and R. Hollingdale), N.Y.: Vintage Books.

Nietzsche, F., 1974, *The Gay Science,* (trans. by. Walter Kaufmann), New York: Random House.

Nietzsche, F., 1990, *Beyond Good and Evil,* Harmondsworth: Penguin.

Nisbet, R., 1969, *Social Change and History,* London: Oxford U. Press.

Nisbett, R. and Cohen, D., 1996, *The Culture of Honor: The Psychology of Violence,* Boulder Colo.: Westview.

Njal's Saga, 1960, in Magnusson, M. and Palsson, H. trans., Harmondsworth: Penguin Books.

Nove, A., 1972, *An Economic History of the Soviet Union,* Harmondsworth: Penguin.

Nunnally, J., 1973, 'Mental Illness: What the Media Present', in Young, 1973.

Orians, G. and Heerwagen, J., 1992, 'Evolved Responses to Landscapes', in, Barkow, Cosmides and Tooby, 1992.

Ornstein, R., (ed.), 1973, *The Nature of Human Consciousness,* San Francisco: W.H. Freeman.

Ornstein, R., 1991, *The Evolution of Consciousness,* New York: Prentice Hall Press,

Panksepp, J., 1982, 'Towards a General Theory of Emotions', *Behavioral and Brain Science,* 5.

Panksepp, J., 1990, 'Gray Zones at the Emotion/Cognition Interface: A Commentary. *Cognition and Emotion,* 4 (3), pp., 289-302.

Panksepp, J., Siviy, S. and Normansell, L., 1985, 'Brain Opioids and Social Emotions', in Retie and Field (eds.), 1985.

Parkin, D., (ed.), 1985, *The Anthropology of Evil,* Oxford: Basil Blackwell.

Parsons, T., 1937, *The Structure of Social Action,* Glencoe: The Free Press.

Parsons, T., 1966, *Societies: Evolutionary and Comparative Perspectives,* Englewood Cliffs, N.J.: Prentice-Hall.

Parsons, T., 1967, *Sociological Theory and Modern Society,* New York: The Free Press.

Passmore, J., 1970, *The Perfectibility of Man,* London: Duckworth.

Peele, S. and Brodsky, A., 1977, *Love and Addiction,* London: Abacus.

Petre-Quadens, O., and Schlag, J. (eds.), 1974, *Basic Sleep Mechanisms,* N.Y.: Academic Press.

Piattelli-Palmarini, M. (ed.), 1980, *Language and Learning: The Debate Between Jean Piaget and Noam Chomsky,* London: Routledge and Kegan Paul.

Pinker, S. and Bloom, P., 1992, 'Natural Language and Natural Selection', in Barkow, Cosmides and Tooby, 1992.

Pinker, S., 1998, *How The Mind Works,* Harmondsworth: Alan Land The Penguin Press.

Planalp, S., 1999, *Communicating Emotion; Social, Moral, and Cultural Processes,* Camb.: Camb. U. Press.

Plant, R., 1991, *Modern Political Thought,* Oxford: Blackwell.

Plato, 1974, *The Republic,* Harmondsworth: Penguin.

Plutchik, R., 1980, *Emotions: A Psychoevolutionary Synthesis,* N.Y.: Harper and Row.

Plutchik, R., 1984, 'Emotions: A General Psychoevolutionary Theory', in Scherer and Ekman, 1984.
Pocock, D., 1985, 'Unruly evil', in Parkin, 1985.
Pope, K., 1980, *On Love and Loving,* San Francisco: Jossey-Bass.
Popper, K., and Eccles, J., 1977, *The Self and Its Brain: An Argument for Interactionism,* N.Y.: Springer International.
Power, M., 1988, 'The Cohesive Foragers: Human and Chimpanzee', in Chance and Omark, 1988.
Rachman, S., 1978, *Fear and Courage,* San Francisco: W.H. Freeman.
Radcliffe-Brown, A. R ., 1952, *Structure and Function in Primitive Societies,* London: Cohen and West.
Radcliffe-Brown, A. R., 1952a, 'On Joking Relationships', in Radcliffe-Brown, 1952.
Radcliffe-Brown, A. R., 1952b, 'Taboo', Radcliffe-Brown, 1952.
Rasmussen, H., 1968, 'Organisation and Control of Endocrine Systems', in Williams, 1968.
Rasmussen, H., 1968a, 'The Parathyroids', in Williams, 1968.
Read, P., 1994, *Ablaze: The Story of Chernobyl,* London: Mandarin.
Reite, M. and Field, T., (eds.), 1985, *The Psychobiology of Attachment and Separation,* Orlando, Fla.: Academic Press.
Rescher, N. 1987, *Scientific Realism: A Critical Reappraisal,* Dordrecht: D. Reidel.
Rhodes, R., 1986, *The Making of the Atomic Bomb,* N.Y.: Simon and Schuster.
Rhodes, R., 1995, *Dark Sun: The Making of the Hydrogen Bomb,* N.Y.: Simon and Schuster.
Richardson, P. and Boyd, R., 1978, 'A Dual Inheritance Model of the Human Evolutionary Process 1: Basic Postulates and a Simple Model', *Jr. Social Biol. Structure* (1).
Ridley, M., 1993, *The Red Queen,* Harmondsworth: Penguin.
Ridley, M., 1999, *Genome,* London: Fourth Estate.
Ridley, Mark,(ed.), 1994, *A Darwin Selection,* London: Fontana.
Ritvo, L., 1990, *Darwin's Influence on Freud: A Tale of Two Sciences,* New Haven: Yale U. Press.
Ritzer, G., 1992, *Sociological Theory,* N.Y.: McGraw-Hill.
Robbins, R., 1959, *The Encyclopedia of Witchcraft and Demonology,* London: Spring Books.
Rosenberg, A., 1981, *Sociobiology and the Preemption of Social Science,* Oxford: Basil Blackwell.
Rosenberg, M., 1979, *Conceiving the Self,* London: Allen and Unwin.
Ross, A. (ed,), 1988, *Universal Abandon: The Politics of Postmodernism,* Minneapolis, U. of Minnesota Press.
Rossi, A., 1977, 'A Biosocial Perspective on Parenting', *Daedalus,* 106:9-31.
Rostow, W., 1971, *The Stages of Economic Growth,* London: Camb.: U. Press.
Roth, P., 1971, *Portnoy's Complaint,* London: Transworld, Corgi Books.
Rothschild, K. (ed.), 1971, *Power in Economics,* Harmondsworth: Penguin.
Rousseau, J., 1968, *The Social Contract,* London: Penguin.
Runciman, W. G., 1966, *Relative Deprivation and Social Justice,* London: Routledge and Kegan Paul.
Runciman, W. G., 1998a, *The Social Animal,* London: Harper-Collins.
Runciman, W. G., Maynard Smith, J. and Dunbar, R. (eds), *Evolution of Social Behaviour Patterns in Primates and Man,* Oxford: Oxford U. Press.
Runciman, W. G., 1998, 'The Selectionist Paradigm and Its Implications for Sociology', *Sociology,* Vol 32 No. 1 (Feb.).
Ruse, M., 1979, *Sociobiology: Sense or Nonsense?,* Netherlands: Dordrecht D. Reidel.
Ruse, M., 1986, *Taking Darwin Seriously,* Oxford: Basil Blackwell.
Russett, C., 1976, *Darwin in America: The Intellectual Response,* San Francisco: Freeman.
Ryan, A., (ed.), 1993, *Justice,* Oxford: Oxford U. Press.
Sagan, C., 1977, *The Dragons of Eden,* N.Y.: Hodder and Stoughton.

Sahlins, M., 1976, *The Use and Abuse of Biology,* Ann Arbor: U. of Michigan Press.
Sahlins, M., 1996, 'The Sadness of Sweetness: The Native Anthropology of Western Cosmology', *Current Anthropology,* 37, pp, 395-428.
Saint-Simon, H., 1964, *Social Organization, The Science of Man, and other Writings,* (F. Markham ed.), New York: Harper Torchbooks.
Salmon, W., 1984, *Scientific Explanation and the Causal Structure of the World,* Princeton, N.J.: Princeton U. Press.
Scherer, K., and Ekman, P., (eds.), 1984, *Approaches to Emotion,* Hillsdale N.J.: Lawrence Erlbaum.
Schneider, D., 1968, 'Some Muddles in the Models: or, How the System Really Works', ` in Banton, 1968.
Schneirla, T., 1957, 'The Concept of Development in Comparative Psychology', in Harris, 1957.
Schneirla, T., 1966, 'Behavioral Development and Comparative Psychology', *Quart. Rev. of Biol. 41: 283-302.*
Schoeck, H., 1970, *Envy: A Theory of Social Behavior,* San Francisco: Harcourt Brace and World.
Schumpeter, J., 1943, *Capitalism, Socialism and Democracy,* London: George Allen and Unwin.
Scientific American, 1992, *Scientific American, Special Issue, Mind and Brain,* 1992 (Vol. 267 No. 3 (Sept.).
Sedman, H., 1970, *Politics, Position and Power,* London: Oxford U. Press.
Seidman, S. and Wagner, D. (eds.), 1990, *Postmodernism and Social Theory,* New York: Basil Blackwell.
Seligman, A., 1992, *The Idea of Civil Society,* New York: The Free Press.
Seligman, M., 1975, *Helplessness: On Depression Development and Death,* San Francisco, Freeman and Co.
Service, E., 1963, *Profiles in Ethnology,* N.Y.: Harper and Row.
Shatz, C., 1992, 'The Developing Brain', *Scientific American,* Vol. 267:3 (Sept.).
Shettleworth, S., 1972, 'Constraints on Learning', *Advances in the Study of Behavior,* 4.
Simons, J., 1995, *Foucault and the Political,* London: Routledge.
Simpson, G., 1961, *Principles of Animal Taxonomy,* New York: Columbia U. Press.
Skidmore, T., 1967, *Politics in Brazil, 1930-1964,* Oxford: Oxford U. Press.
Sloan, A., 1972, *My Years with General Motors,* N.Y.: Anchor Books.
Smith, A., 1981, *An Inquiry into the Nature and Causes of the Wealth of Nations,* Indianapolis: Liberty Press.
Somit, A., 1981, 'Human Nature as the Central Issue in Political Philosophy', in White, 1981.
Sorokin, P., 1947, 'What is A Social Class?', *Jr. of Legal and Political Sociology',* (1947), pp., 21-28.
Spencer, H., 1862, *First Principles,* London: Williams and Norgate.
Spencer, H., 1876, *The Principles of Sociology,* London: Williams and Norgate.
Spencer, H., 1891, 'Progress: Its Laws and Causes', in, *Essays, Scientific, Political and Speculative,* (3 Vol), London: Williams and Norgate.
Spencer, H., 1967, *The Evolution of Society: Selections From Herbert Spencer's Principles of Sociology,* Chicago: U. of Chicago Press.
Springer S. and Deutsch, G., 1981, *Left Brain, Right Brain,* N.Y.: Freeman.
Stein, E. (ed.), 1992, *Forms of Desire, Sexual Orientations and the Social Constructionist Controversy,* London: Routledge.
Steiner, F., 1967, *Taboo,* Harmondsworth: Penguin Books.
Stepan, A., 1971, *The Military in Politics: Changing Patterns in Brazil,* Princeton: Princeton U. Press.
Stirk, P. and Weigall, D., 1995, *An Introduction to Political Ideas,* London: Printer Pub.
Stockman, D., 1986, *The Triumph of Politics,* N.Y.: Hodder and Stoughton.
Strongman, K., 1973, *The Psychology of Emotions,* N.Y.: John Wiley.

Sumner, W. G., 1992, *On Liberty, Society and Politics,* Flint Michigan: The Liberty Press.
Sweeney, J., 1997, 'Politics: Labour After Brighton: Laugh if you Dare. There's a Serious Side to Foetusgate', *The Observer,* (Sunday, Oct. 5), news page 24.
Symons, D., 1979, *The Evolution of Human Sexuality,* London: Oxford U. Press.
Symons, D., 1992, 'On the use and Misuse of Darwinism in the Study of Behavior', in Barkow, Cosmides and Tooby, 1992.
Tanner, N., 1981, *On Becoming Human,* Camb.: Cambridge U. Press.
Tate, C., 1989, 'In the 1800s Anti-Smoking was a Burning Issue', *Smithsonian,* 20: 107-117.
Taylor, S. and Brown, J., 1988, 'Illusions and Well Being: A Social Psychological Perspective on Mental Health', *Psychological Bulletin,* 103,, pp., 193-210.
Tealby, N., 1928, *Candide and Other Romances,* (trans. Richard Aldington), London: John Lane The Bodley Head.
Thody, P., 1997, *'Don't Do It! A Dictionary of the Forbidden,* London: Athlone Press.
Thorpe, W., 1979, *The Origin and Rise of Ethology* , London: Heinemann.
Tooby, J and Cosmides, L., 1992, 'The Psychological Foundations of Culture', in Barkow, Cosmides and Tooby, 1992.
Trigg, R., 1970, *Pain and Emotion,* Oxford: Clarendon Press.
Trigg, R., 1982, *The Shaping of Man: Philosophical Aspect of Sociobiology,* Oxford: Basil Blackwell.
Trivers, R., 1971, 'The Evolution of Reciprocal Altruism', *Quarterly Review of Biology,* 46:35-57.
Trivers, R., 1972, 'Parental Investment and Sexual Selection', in Campbell, B., 1972.
Trivers, R., 1981, 'Sociobiology and Politics', in White, 1981.
Trivers, R., 1985, *Social Evolution,* Menlo Park, Calif.: Benjamin/Cummings Pub. Co.
Turke, P. and Betzig, L., 1985, 'Those Who Can Do: Wealth, Status, and Reproductive Success on Ifaluk', *Ethology and Sociobiology,* 6:2 .
Turnbull, C., 1968, 'The Importance of Flux in Two Hunting Societies', in Lee and DeVore, 1968.
Turnbull, C., 1976, *The Forest People,* London: Pan Books.
Turner, B. 1993, *Citizenship and Social Theory,* London: Sage.
Twitchell, J. 1987, *Forbidden Partners: The Incest Taboo in Modern Culture,* New York: Columbia U. Press.
Van den Berghe, P., 1979, *Human Family Systems,* London: Elsevier.
Van Sommers, P., 1988, *Jealousy: What is it and Who Feels it?,* Harmondsworth: Penguin.
Vellacott, P., 1956, *Aeschylus: The Oresteian Trilogy,* Harmondsworth: Penguin.
Voltaire, 1747, 'Memnon' in Tealby, 1928.
Voltaire, 1759, *Candide,* in Tealby, 1928.
Voltaire, 1963, *The Age of Louis XIV and Other Selected Writings,* N.Y.: Washington Square Press.
Voltaire, 1972, *Philosophical Dictionary'* Harmondsworth: Penguin.
Voltaire, 1980 *Letters on England,* Harmondsworth, Penguin.
Walter, E., 1969, *Terror and Resistance: A Study in Political Violence,* Oxford: Oxford U. Press.
Ward, L. 1893, *Psychic Factors in Civilization,* Boston: Ginn.
Ward, L., 1897 *Dynamic Sociology,* N.Y.: Appleton.
Ward, L., 1906, *Applied Sociology,* Boston: Ginn.
Warner, W. L., 1960, *Social Class in America: A Manual of Procedure for the Measurement of Social Status,* (with Marchia Meeker and Kenneth Eells), N.Y.: Harper.
Warner, W. L. and Lunt, P., 1942, *The Status System of A Modern Community,* New Haven: Yale U. Press.
Warner, W. L.,. and Lunt, P., 1941, *The Social Life of a Modern Community,* New Haven: Yale U. Press.
Washburn, S. and McCown, E. (eds), 1978, *Human Evolution: Biosocial Perspectives,* Menlo Park, Calif: Benjamin/Cummings.

Washburn, S., 1963, *Classification and Human Evolution,* Chicago: Aldine.
Washburn, S., 1963a 'Behavior and Human Evolution', in Washburn, 1963.
Watzlawick, P., 1978, *The Language of Change: Elements of Therapeutic Communication,* N.Y.: Basic Books.
Webb, W. (ed.), 1973, *Sleep: an Active Process,* Glenview, Ill., Scott, Foresman.
Weber, M., 1978, *Economy and Society,* Vol I. Berkeley: U. of California Press.
Weber, M., 1981, *General Economic History,* New Brunswick: Transaction Books.
White, E., 1981, *Sociobiology and Human Politics,* Lexington Mass.: Lexington Books.
White, L., 1949, *The Science of Culture,* New York: Grove Press.
Whiten, A., 1997, 'The Machiavellian Mindreader', in Whiten and Byrne, 1997.
Whiten, A., and Byrne, W., 1997, *Machiavellian Intelligence II: Extensions and Evaluations,* Camb.: U. Of Camb. Press.
Willhoite, F., 1981, 'Rank and Reciprocity: Speculation on Human Emotions and Political Life', in White, 1981.
Williams, G. C., 1966, *Adaptation and natural selection: A Critique of some current Evolutionary Thought,* Princeton, N.J.: Princeton University Press.
Williams, R., 1968, *Textbook of Endocrinology,* London: Saunders.
Willis, R., 1970, 'The Kamcape Movement', in Marwick, M., 1970.
Willis, R., 1970a, 'Instant Millennium: The Sociology of African Witch-Cleansing Cults', in Douglas, 1970a.
Wilson, M. and Daly, M., 1992, 'The Man who Mistook his Wife for a Chattel', in Barkow, Cosmides and Tooby, 1992.
Wilson, B., 1990, *The Social Dimensions of Sectarianism, Sects and New Religious Movements in Contemporary Society,* Oxford: Clarendon.
Wilson, E. O., 1975, *Sociobiology: The new Synthesis,* Camb. Mass.: The Belknap Press.
Wolfe, T., 1980, *The Right Stuff,* N.Y.: Bantam Books.
Woodburn, J., 1982, 'Egalitarian societies', *Man,* 17, 431-451.
Young, A., 1996, *Imagining Crime,* London: Sage.
Young, J., 1973, *The Manufacture of News,* London: Constable.
Zahavi, A., 1975, 'Mate Selection - a Selection for a Handicap', *Jr. of Theoretical Biology,* 53, pp., 205-214.
Zald, M. (ed.), 1970, *Power in Organizations,* Nashville Tenn.: Vanderbilt U. Press.
Zeki, S., 1992, 'The Visual Image in Mind and Brain', *Scientific American,* Vol. 267:3 (Sept.).
Zihlman, A., 1981, 'Women as Shapers of Human Adaptation', in Dahlberg, 1981.
Zihlman, A., 1983, *The Human Evolution Coloring Book,* N.Y.: Barns and Nobel.

INDEX